MEMORIES OF THE

Leicestershire Coalfields

David Bell

COUNTRYSIDE BOOKS
NEWBURY BERKSHIRE

First published 2007
© David Bell, 2007

COUNTRYSIDE BOOKS
3 Catherine Road
Newbury, Berkshire

To view our complete range of books,
please visit us at
www.countrysidebooks.co.uk

ISBN 978 1 84674 061 9

Designed by Peter Davies, Nautilus Design

Produced through MRM Associates Ltd., Reading
Typeset by CJWT Solutions, St Helens
Printed by Borcombe SP Ltd., Romsey

*All material for the manufacture of this book
was sourced from sustainable forests.*

Contents

Acknowledgements

would like to thank the following people for the help they gave me: Denis Baker, Dave Barker, Kathy Belfield, Michael J. Conibear, Frank Gregory, Barrie Hall, Ken Hillier, Derrick Holmes, Derek Howe, Robert Jones, Chris Matchett, Paul Monk, Cecil 'Ozzy' Osborne, Maureen Partridge, Alan Pearson, Alan Ratcliffe, Mick 'Richo' Richmond, Peter Smith, Malcolm Tudor, Ken Ward and Joe White.

I would also like to acknowledge the help from Whitwick Historical Group, Desford Brass Band, Ashby Museum, Ashby Image and Print, Ashby Tourist Information Centre, Ashby Library, and Magic Attic archives in Swadlincote.

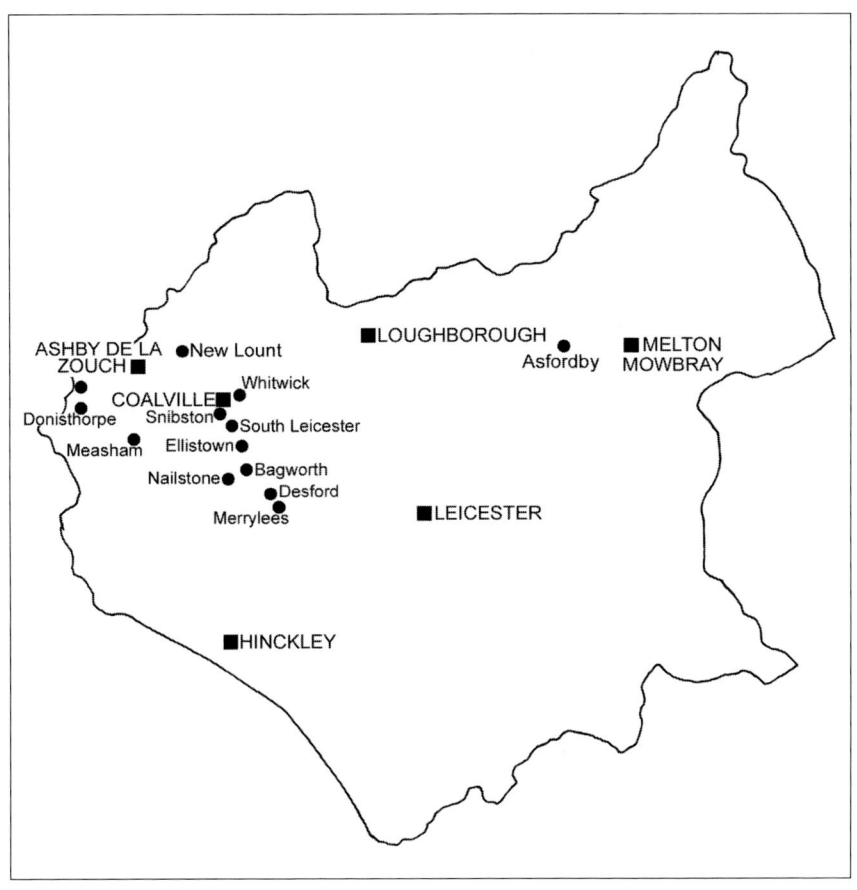

Michael Gould from Thringstone with Neal, his pit pony (see Chapter 3).

Introduction

A **centuries-old way of life** came to an end in Leicestershire when Bagworth Colliery, the last of the traditional deep coal mines, closed in 1991. Many of the men who earned their livelihood in the mines of Leicestershire have been eager to share their experiences, and it is their memories of the work, the dangers, the humour and the comradeship of the mining communities that form the basis of this book.

Leicestershire has been mined for coal for centuries and in the 1800s there were many collieries with rather exotic names: California, Calcutta, Alabama. There was also one named Califat, where there was a disaster in 1863 when a wall into the old Limby Hall pit was breached and the inrush of water drowned three men.

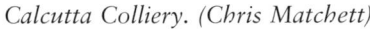

Calcutta Colliery. (Chris Matchett)

Even some of the collieries that were simply named after the villages where they were situated, had local nicknames. New Coleorton Colliery was always known as 'Bug and Wink'. This strange appellation has puzzled historians for decades, but one possible explanation emerged in a magazine called *Snibbets*, produced for the miners at Snibston pit. The suggestion was that in the days of private ownership of the mines, the owners of New Coleorton were crafty negotiators when dealing with the men. The miners claimed that the bosses there were always able to humbug and hoodwink them, and thus the pit became known as 'Bug and Wink'. Bug and Wink pit was in use from 1875 until 1918.

In the late 1950s, there were still nine working collieries in the north-west Leicestershire coalfield, namely Bagworth, Desford, Ellistown, Merry Lees, Nailstone, New Lount, Snibston, South Leicester and Whitwick. The pits fell into two geographical groups. One was based in and around Coalville, which included Whitwick, Snibston, South Leicester and New Lount, while the other group was clustered around Bagworth (strangers to the area should note that 95% of residents of Bagworth pronounce it without the *w* – Baggerth).

Collieries at Donisthorpe, Moira (Rawdon and Marquis) and Measham (Measham and Minorca) all fall within the county of Leicestershire, but are regarded as belonging to South Derbyshire coalfield. This is because the boundary between South Derbyshire coalfield and the North West Leicestershire coalfield follows an underground feature known as the Boothorpe fault, rather than the county border. Recollections of those pits have been recalled in my previous book, *Memories of the Derbyshire Coalfields*.

Some pit names can be misleading. Desford colliery, for instance, is not in the village of Desford, but within Bagworth parish. The most misleading of all is South Leicester – shortened by most miners to simply 'South'. South Leicester colliery is nowhere near Leicester, and it is certainly not south of the city, but lies just to the south of the Whitwick and Snibston pits.

The oldest of these nine pits were Whitwick, New Lount and

Bagworth, sunk in 1820, 1824 and 1828 respectively. Whitwick had three shafts that were sunk in 1820, another in 1874 and a fifth in 1901. The founder of the colliery was William Stenson, a native of Coleorton, who had gained experience of mining engineering in Derbyshire and Gloucestershire, before coming back to his native county to sink the mine at Whitwick.

Snibston had an illustrious founder. In 1831, George Stephenson, the railway pioneer, bought the Snibston estate and immediately began to sink the first shaft of what became Snibston colliery. Together with his son Robert, he had already built a railway to transport coal from Swannington to Leicester, and this line passed close by the new mine. George sank a second pair of shafts off a nearby country road called Long Lane, and these formed the basis of the new colliery, which began to produce coal in 1832. Miners came to the area from County Durham, and they were housed first in tenements, then in terraced cottages on Long Lane. Later, a school was built and shops opened. The settlement around Snibston grew into a town, known as Coalville. Another shaft at Snibston Colliery was added in 1914, and a drift in 1963.

Residents of Coalville today are proud to acknowledge their illustrious founder, and the town boasts a Stephenson College and a bypass named George Stephenson Way. Coalville is, of course, much younger than the surrounding villages, celebrating its comparatively youthful 175th anniversary in 2007. It had some unusual features, one of them a level crossing in the High Street that frequently closed to allow goods trains to cross. The coal mine was located in the middle of the town. Passers-by and shoppers could see scores of miners, covered in coal dust, crossing the main road at the end of a shift to the pit baths on the other side.

It may have been Durham miners who came to Coalville in the 1830s, but in the 1960s it was miners from Scotland who came to work in the Leicestershire pits. Many of them settled in the village of Thringstone, which is why the biggest centre of social activity in that village was the Rangers Supporters' Club, named after the Glasgow football team.

Headstock at Snibston pit. (Michael Conibear)

Coleorton pit, 1927. (Michael Conibear)

When the coal industry was nationalised in 1947, Coleorton Hall became the regional administrative headquarters of the National Coal Board. This Grade II listed mansion house was once the home of the Beaumont family, and a centre of the arts – Sir George Beaumont entertained the poets Byron, Southey, Constable, Wordsworth and Coleridge there. Sir Walter Scott stayed at the Hall, and used the nearby town of Ashby-de-la-Zouch as the setting for the tournament in his novel *Ivanhoe*. Sir George was one of the founders of the National Gallery, and had a memorial to Sir Joshua Reynolds built in the grounds of the Hall.

Initially covering the Leicestershire and Derbyshire coalfields, the NCB region eventually took in Warwickshire, Kent and part of Staffordshire. Ken Ward, who worked there as an archivist, tells me that there were approximately 450 staff, including area managers, surveyors, geologists, specialist engineers, draughts-men, photographers, estate managers, contract purchasers, and departments dealing with sales and industrial relations.

Unlike Coalville and its surrounding villages, Ashby-de-la-Zouch was a town with no mines underneath it, and therefore the town suffered no subsidence problems. Maybe it was not entirely coincidental that many Coleorton Hall staff decided to make their homes there. Ashby was also the location of the Mines Rescue Service, though that has recently moved to the National Forest centre near Moira.

The pits at New Lount, Nailstone and Merry Lees closed down in the 1960s, the latter because of trouble with flooding. It suffered from a feature miners refer to as 'Red Measures'. Frank Gregory, a miner at Desford pit, the nearest neighbour to Merry Lees, explained, 'There were many water problems in the pits round here. Underneath Merry Lees, there was an underground river and an underground lake called Red Measures. Lakes underground are not just like water, the pervious rock soaks it up like chalk, so that's the river but it's soaked up. It's there but it moves.' Alan Ratcliffe, a mining surveyor, put it slightly differently, describing Red Measures as water-bearing sandstone. Under either description, it was a real problem at Merry Lees, and the mine closed, although for a while its coal was reached from Desford pit.

Although Nailstone closed as a pit, the screens where the coal was sorted remained, and a drift was built so that coal from Bagworth, Desford and Ellistown could be transported underground to the surface at Nailstone, to be graded. Frank Gregory explained, 'Nailstone pit was closed in 1967, but after that it was still alive, because they kept it open for washing and grading the coal – screening – for the whole area. From Bagworth, they made a special underground roadway, graded it up to the surface and it came up at Nailstone on the pit site. Coal came from Ellistown and from Desford to Bagworth and then it all went up that drift to Nailstone. And that stayed open until all the pits closed, until the last pit closed, and that was Bagworth.'

Snibston Colliery and Desford actually closed down in 1984, during the miners' strike that was trying to keep pits open. South

Leicester and Whitwick followed in 1986, leaving Ellistown and Bagworth as Leicestershire's last two pits. Ellistown closed in 1990 and Bagworth a year later. And so, it might be thought, coal mining in Leicestershire came to an end.

But it didn't. A whole new coalfield was discovered in a different area of the county. This was in north-east Leicestershire, under the Vale of Belvoir, where a good-sized coalfield with high grade coal appeared to be a lifesaver for the miners whose pits were closing. The original plan was for three collieries: one in the north of the Vale of Belvoir, one in the centre near the villages of Hose, Long Clawson and Harby, and one in the south, two or three miles west of Melton Mowbray.

However, the enthusiasm of the miners was not shared by all the residents of the Vale of Belvoir. One of the main objectors to any mining in the Vale was the late Duke of Rutland. Not only would mining ruin the views from his home, Belvoir Castle, it would change the nature of the rural farmland. Who would want to work on the land when they could earn much more down the new pits? The Duke was a formidable opponent. Not only was he the feudal lord of the area, he was also its county councillor. The battle over mining in the vale became heated, even intemperate. Not only would the coming of mining spoil the views, it would bring in that alien species – the miner. This prospect was regarded by some with fear and hostility. The stereotypical view was that miners were rough, they drank, they had strange hobbies, they voted the wrong way, and they were probably poachers. They were not wanted.

There was one note of wry humour. When the Duke announced that if mining was allowed in the vale, he would lie down in front of the first bulldozer, 37 Coalville miners volunteered to drive it. A mock raffle was even held in the Miners' Welfare, the only prize being the right to drive that bulldozer. Alan Pearson, a miner from Snibston pit, added, 'With me being on the council, I actually went to tea with the Duke of Rutland. The conversation got round to the Vale of Belvoir, and I said to him, "I bet if they offered you six pence a ton, you wouldn't be lying in front of the first bulldozer, would you?" And he looked at me with his one eye and he says, "What do you do for a living?" and I said I worked at the pit. I could see his hackles go up, but to be honest we got on fine.'

'Ozzy' Osborne, a shaftsman from Ellistown pit, told me of a visit he made to Belvoir Castle at the time. 'When they were going to build the pit at Asfordby, the Duke of Rutland didn't want pits or miners near Belvoir Castle. Me and the wife and a couple of friends went to the castle on a trip, and they'd got posters up, saying "Keep Out – No Coal Mines". And on the desks they'd got these petitions for visitors to sign against the coal mining. Well, by the big fireplaces in all the rooms they'd got these big lumps of coal, so I says, "What's that stuff, down there?" And the young lady said, "It's coal." I says, "Is it now? Well, where does it come from?" She said, "Oh, Lord so-and-so has his own mines. It comes from there." "Ooh, yes," I says, "you don't want us to get the local coal then?" Well, she didn't know what to say. Everybody was looking and listening, and they realised I were a miner and I were pulling her leg. In the castle kitchen there were great tubs full of coal, and I says, "What's this stuff in here? I've got some tubs of coal just like these." They said, "Have you? Where?" I says, "Down the pit at Ellistown." They looked at me gone out. I thought they were going to tell me off but they didn't. I was just stirring it up.'

Despite the need for coal, and despite the need for jobs, the Duke of Rutland won the battle. The colliery in the north and the colliery in the centre of the Vale of Belvoir were never built. The revised plan was for just one pit on the very southern edge of the coalfield, at Asfordby Hill, about three miles west of Melton Mowbray. There was already an iron foundry there, Holwell Works, and many of the local residents worked there.

That is where the pit – Asfordby Colliery – was sunk. However, the miners encountered impossible geological conditions. Joe White, who had worked at Ellistown and Bagworth, and then transferred to the new pit at Asfordby, told mc, 'Asfordby was mainly fraught with geological bad planning. It was in the wrong place. I actually started on big developments which were classed

Asfordby miners. (Chris Matchett)

as civil engineering really, massive crossings which we hoped, once the shaft was sunk, to divert into different roadways. However, there were internal combustions into different coal seams, which caused underground fires.'

Joe also told me of terrible roof falls, where it was a miracle that no one was hurt. 'To be fair,' Joe continued, 'the British Coal and NCB management were aware of the geological mistakes and made them known at the time, but no one listened, it was just brushed aside. Those coal seams were viable, but not from Asfordby.'

Asfordby Colliery began producing coal in 1994 and closed in 1997. It was a three-year wonder. Derek Howe, a miner at Snibston, put it quite simply: 'That pit at Asfordby only lasted a few years, it was built in the wrong place, it should have been at Harby, that's where the coal was.'

The coal was there, under the Vale of Belvoir, and it still is. As Joe White explained, 'In geological terms, in mining terms, all of the reserves in the Vale of Belvoir coalfield, in Leicestershire, for

Asfordby mine – work started on the site in August 1984, with the first coal expected in 1992 and full production by 1993. (Chris Matchett)

Leicestershire miners, are still there today. They are still viable seams of low sulphur coal, coal that would be a good supply to the energy market. I do fervently believe that.'

So Leicestershire ceased to be a coal mining area, except for a number of open cast sites, universally hated by the people who live near them. They are quarries rather than mines. Now, if in the future, open cast mining ever comes to the Vale of Belvoir, the residents may perhaps wish that they had opted for the traditional deep mines.

Chapter 1

Cages and Winding Wheels

Now that every pit in Leicestershire has closed, the iconic sight of a pit head with its winding wheel has virtually disappeared. It was the part of the mine that the general public would see as they travelled past, and could be as exciting and wonderful as a windmill or a castle. Even to anyone from a mining area, to see the wheel going round was something special, something majestic.

Alan Ratcliffe qualified as a mining surveyor after starting work down the pit at Whitwick, and was one of the men who had to climb up the headgear, as part of his job. 'We'd be up there for sighting purposes, you see, to measure the headgear and take observations to surrounding high points, like Bardon Hill and the headgear of other nearby collieries. Now going up that headgear was hairy the first time. Because in the early days – I'm talking about the 1940s – the outside legs of the headgear sloped inwards, so when you were climbing up them there was a lean. And that's the side where the handrails were. That's all you'd got, a handrail. There were no rails both sides, no latticework fencing or chainlink fencing like you see festooned round headgear now.

Alan Ratcliffe (Area Surveyor) and Trevor Taylor (National Chief Surveyor) at Bagworth Colliery. (Alan Ratcliffe)

'It was a bit of a climb when you'd got a theodolite structure on your shoulder or a set of "legs" under your arm, leaning inwards on these railings, and seeing the ground disappear down below. But it was something you got used to. The last bit of the climb, you went up a vertical ladder and you'd be covered in grease by that time. You came through a hole in the flat top, and then you were on the level. That was where the axles of the winding wheel were. There were guardrails round the side, but it was just a stanchion and two rails. You kept well clear of the wheel, because it would still be turning and the whole platform would be shaking. It was quite dangerous, but nobody stopped us doing it, there was no Health and Safety then.'

Those winding wheels pulled up and lowered the cages that took the miners down the shaft into the mine itself. The shafts were the connection between two worlds, the surface world of sunshine and daylight and the underground world where the men actually worked at getting the coal – *turning coal*, as they called it. There were a few mines that had a sloping entrance, known as a drift, into them – Merry Lees was one. Snibston itself, the colliery in the heart of Coalville, had a drift, built in 1963, to supplement the much older shafts. Alan Pearson recalled, 'When you were walking up the drift at Snibston, which was 918 yards long, and a one in four gradient, you could see the light at the top, and you were always watching that light.'

Most of the pits, however, were entered down the shaft, in a cage. Alan Ratcliffe described what it was like to go down. 'When the cage was at rest on the pit top, beneath it, in the walls of the shaft, were two sprags that would come out. The cage would sit on them. So before it could go down again, the engine winder would have to lift the cage up a few inches, and the sprags would be retracted into the wall. When you were on the cage, particularly in the early morning, when you were going down at half past six or at seven o'clock, the buzzer blew. If you weren't there you'd had it. They wouldn't let you down.

'Now the winder would be rushing because he'd got a queue to get rid of before seven, so you'd get on the cage, packed in with

The drift at Merry Lees pit. (Michael Conibear)

the men. You'd feel it rise, then zonk, down you'd go. You should have heard the swearing. But over the years, they stopped that. They fitted governors to the winding engine, so if they tried to go too fast it stopped. Now that's an unpleasant experience. I've had that, where he's exceeding his speed and the governors suddenly kick in and stop it dead like a brake going on. And you're bouncing up and down. And you think, oh dear, or words to that effect.

'The cage settles down then and off you go again. But when you're bouncing up and down, it's not very nice. We used to jest about it dropping all the way. We used to say, "I've got the perfect solution. Just before you hit the bottom, jump up. You'll be in mid-air then when it hits the bottom." But your life was in the winder's hands. It didn't matter how hardened the pitmen were,

Bagworth headstock. (Chris Matchett)

it'd always turn your stomach over. It was like on fairground rides and the Tower of Terror at Disneyworld.

'I'd already had a trip down the pit at Whitwick when we went down the old disaster shaft, shaft number five. There was a collapse there, in 1941. It still had its wooden headgear, as at the time of the disaster, and it had still got a flat winding rope, like khaki webbing. You could see through it. It was about three or four inches wide but absolutely flat. And as you stood at the cage – it was a small one, because the shaft was only 8 ft in diameter – you could see the rope stretching up above and you could see through it, through the weave. It was my first trip ever, and I was apprehensive but not nervous. There were quite a few of us, perhaps half a dozen, all office staff. One of the under-managers took us down. Of course the winder took it a bit fast, they always did with new people on. It was great. I was at home, down there, the smell and the dark. It was no problem.'

'Ozzy' Osborne first went down the pit when he was nine, when the engine winder at Nailstone took him and the landlord's son from the Corner Pin pub. 'He took us on a Sunday morning, when the pit wasn't working. We were just taken down to the bottom and a little way outbye to see the horses, the ponies. The cage went down very steady. Normally with visitors on, especially if there's some women, they dropped it really fast.'

Some young lads were apprehensive about their first trip down in the cage. Frank Gregory said, 'When I was at school, I went on a trip down Whitwick Colliery, and I said to myself when we got down there, no way will I ever work down this. It was terrifying. And we only went round the pit bottom with the lights. We didn't realise there were no lights further on, only your cap lamp. So that were my first taste of it.' Alan Pearson also admitted, 'The first time down the cage felt very strange.'

This apprehension was not surprising: the cage could be a dangerous place. Derek Howe, who became a workman's inspector for the NUM, told me, 'Snibston had had a cage accident before I went. You could tell the blokes who'd been in it because they were still limping. That were the last cage accident they had at Snibston. All the lads wanted a fag on the pit top before they went down again. Of course they were pit bottom blokes, there were very few face workers. There were lots hurt, but fortunately no one died.'

However, not everyone was dreading the trip down. For some it wasn't just a short journey, it was an important rite of passage. According to Derrick Holmes of Bagworth, 'My first trip down the cage, I was really up for it. Because you see, I'd often been on the pit bank, on the surface, as a kid. That was my playground, my back yard. We used to play there, so that was nothing special. I was looking forward to going down in the cage, so I was never nervous, not the slightest bit. You thought once you got down underground, you were a man.'

Malcolm Tudor grew up near Ellistown. 'The first time I went down the shaft, I was only a school kid. My father took me down Ellistown Colliery, with permission, and we went down one Saturday afternoon. When we dropped, I thought somebody had cut the rope. It was steam winding and so it was very quick. It was the same when I started working at South Leicester, it was steam

Ellistown Colliery. (Chris Matchett)

winding, and you felt like your stomach was going up through your neck. But I soon found out if you hung on to the handrail at the side of the cage and leaned forward slightly it didn't affect you.

'What they were worried about when they took the steam winding out and put the electric in, was not getting the same amount of coal out, because the electric winding went down slower. But with electric winding, the average speed was better because there was a quicker turn round. When they were taking men up or down in the cage, it would go fast in the middle, but when you came to either end there was a slow banker, especially with the steam winder. It went very slow the last few feet.

'The funny thing was, when you came to the bottom, as it was slowing down you'd feel as if you were going back up. That's because, as it was slowing down, the weight came on your feet. But if you looked out of the cage end, you could see you were still going down.'

The winding man had a responsible job, with the lives of the men in the cage depending on him, but he could occasionally get it wrong. Derek Howe described one really frightening moment as he rode down the shaft. 'I remember one day, going to work down the old shaft at Snibston, and they hadn't got a bottom in the old shaft. And that particular day, we went straight by the men who were waiting to come up. I always remember a chap named Harry Smith; he tried to open the gate when it were going by. If he'd managed it, that would have been someone hurt. The cage went down out of sight. It really frightened us.

'And I remember another day coming up, when it were a late stop – what we called a "doddy". The belts hadn't been going very well and so we were late finishing. We got on the cage and it went down instead of up. We dropped about three lengths of the cage, right down to the water. Obviously it stopped when we hit the water. Oooh, my goodness! When we did get up to the top, the lads all went running for the winder. It really frightened us. And that happened to us twice while I was on. It's a terrible feeling. They did put a bottom in eventually, so it couldn't go further down.'

Barrie Hall, who worked at Snibston pit, reminisced, 'We'd got

Tommy Parsons, engine winder at Whitwick.
(Whitwick Historical Group)

a bloke who used to drive the cage – the engine winder – called Locky, and if he was in a bad mood, he used to drop you. You'd hit the bottom, and your tummy would be about two or three seconds afterwards. When we were going for records and things like that, the cage used to bounce, bang, and up it'd go again.' The cage was also used for the tubs of coal. 'There was one time when I was working in the pit bottom, when we'd got a "danny" – it's a bit longer than a tub but not as long as two tubs – so you'd just have one danny on the cage, but you have to get in and chuck it off the road, so that it wouldn't move. And while the onsetter was doing this, the cage went up without a signal. He was effing and blinding and so on. It took him so far up, until he pressed the red button to stop it.'

Despite the occasional errors, the winding man was often a

25

The engine winder at Lount pit. (Michael Conibear)

character in his own right. The engine room was not just his place of work; it was his domain, his castle. Alan Pearson, who worked at Snibston pit, remembers one of them vividly: 'I'll tell you one of the strangest situations there – only being allowed to stand *outside* the winding house at Snibston. Aubrey Peace who was there at the time, he used to have it spotless, and you could never go inside. He wouldn't let anyone in. And I'm not kidding, you could eat your food off the floor. You really could. All the brass and copper were polished up, and the big winding wheel was painted immaculately red and green. It sticks in your mind, things like that. Aubrey used to have his boiler suit on and, before the invention of the NCB orange overalls, his blue overall suit.'

The trip inside the cage was one thing. Going down while standing on top of the cage was a much more scary ride. 'Ozzy' Osborne was a plumber at Ellistown Colliery, and then went to

Birch Coppice for a fortnight to do his underground training. Back at Ellistown he started helping out the fitters underground with pipework. 'Then,' he explained, 'the shaftsman's job came vacant and I applied for that, and I got it. The shaftsman examines and maintains the shaft. Every day, I had to go up and down all the shafts – we had three shafts. That's your first job. You had to ride and fix any things you find. That's ride on top of the cage, not inside it. You can see more on the top, and you can stride over and see to anything.

'I used to have a big metal buffer off a loco. When I wanted the cage to stop, I didn't shout, I banged this buffer. One bang for stop. The pit top man would signal the winder and then he'd stop the cage. Nine times out of ten, I'd got to come back up a little way and have another look. It didn't bother me, riding on top of the cage. I'd been used to climbing up churches and things when I was a plumber, doing the lead on church roofs. The shaftsman's job also included changing the winding ropes, and doing the brickwork. One shaft would be hot and dusty, the next would be freezing cold.'

'Ozzy' was obviously quite used to riding on top of the cage every day, but for a man who had to do it very occasionally, it was a different matter. Derek Howe, the workman's inspector, told me, 'When it came to going down the shaft on top of the cage, with a safety harness on, even though I'd got gloves on, the chains were embedded in my hands from hanging on so tight. But we had to do it. That was the first time I'd rode on top of the cage.'

Another person who did the same was the surveyor. Alan Ratcliffe explained, 'I got a lot of experience at Bagworth. We went all over the pit as a surveyor, we had experiences that a miner would never have. For instance, shaft work. Riding down on top of the cage. The ordinary miner never got on the top of a cage at all. But it wasn't unusual for us. We had to do it to measure the shaft for one thing. I never had a problem with being on top of the cage. When you got on, you were all right at the side where the shaft wall was, but when you turned round you'd look into space. It was a 22 ft shaft at Bagworth, so you'd be perhaps a couple of feet away in the middle, then you'd got the width of the cage, about 4 ft, and over there you'd got nothing, just a void. And when you got to the meeting place, where the other cage passed, well you always had a shaftsman with you on top, and their signal was a hammer and a great big plate of iron about 2 ft in diameter with a hook in it and it hung on the chains. And that was the only way he could signal. Bang. And the pit top man would respond to that.

'I've seen a shaftman at the meeting of the cages, get over from one cage onto the other. And they wouldn't always have their harness on, they'd swing round these bull chains. Familiarity breeding contempt. Then we had skips installed at Bagworth that were just coal carriers. You didn't ride in a skip, it was purely a coal carrier. It had a sloping floor to it, so that when it got up to the pit top, it just discharged the coal. We did a survey on top of the skip once, and that had just got a single rope, no bull chains, no nothing. I got hold of the rope and hung on. My feet never moved from top to bottom. When we got in the bottom, we heard a bang at the pit top. Everybody just ducked, but nothing came down. We'd thought something was on the way. And you'd no protection at all.

'That was one end of the shaft. At the other end – the bottom end – you landed on huge boards, and the guide ropes went through the boards. And underneath, they had great big cheese weights hanging on them – shaped like a cheese with a slot in it. And of course all the detritus, the debris and water running down the walls – most shafts were wet – it'd make sludge in the bottom, red ochre. Somewhere off the roadway there'd be a sloping road so you could go down and inspect the cheese weights and the bottom of the cables. They'd have to be changed every so often. We did one once in Bagworth shaft, down in the bottom there, and it was thick. We'd got wellingtons on instead of pit boots. The red ochre and brown sludge had covered the boards, so it was filthy. I went to step back but my boots didn't want to go, and I sat in it. And I came out with a great big brown patch on my backside. I got some comments; you can guess what they said.'

Chapter 2

'You're Not Going Down the Pit!'

Miners from other coalfields frequently talk about starting work in the pit at the age of fifteen, straight from school. What struck me when talking to Leicestershire miners was how many of them had not made that immediate, young transition. Even those from mining families went into other occupations for a while.

Although a miner's son, Joe White waited until he was eighteen before starting on his mining career. 'I grew up in Whitwick, in the council houses on Hall Lane. My dad, Tom, was a miner for 40-plus years, at Bagworth and at South Leicester. The whole family were miners. I had a choice of career. I wanted to go in the Navy but Dad talked me out of that. I said, "Well, I'll go down the pit then, Dad." He said, "There's enough idiots down there without another one." Only joking, obviously. He said, "You'd

Whitwick – 15m head. (Whitwick Historical Group)

do a lot better to go into something else." I went paint spraying, but a lot of my mates were in the mining industry or the building trades. So at eighteen, I went down the pit. It was partly because of the money but also the family tradition.

'When I was only fourteen or fifteen, my first trip down the shaft was at Snibston, on a school trip. Until then I could never understand why, when my dad finished work, he'd come home and have his dinner, then he was tired and just nodded off on the settee. When I went down Snibston Colliery we had two or three hours down there because we went right down to the face. We had a shower the same as everybody else, took part in the whole day, which was fantastic. I came home and I remember after my dad had his dinner he was sitting there on the settee, dozing off due to lack of oxygen, toxic gases, etcetera. I was actually sitting at the side of him doing the same. I suppose some people nowadays would say that the older pits were quite scary but I was not daunted by that sort of thing.'

Derek Howe had six years on a farm, before going to work as a miner. 'At fourteen, I started to work at Elm Farm in Swannington, helping to make ice cream, selling milk and generally helping on the farm. I went down the pit when I was 20. I went to Whitwick Colliery when they were asking for miners. I did my training there under a bloke called Jess Holmes. He was in charge, and a very good man. Then I went to Snibston, and I was put on the roads. I worked at Snibston for 35 years.'

'Ozzy' Osborne told me, 'When I left school I went plumbing, as an apprentice at Orton and Company. Funnily enough, we did a lot of work at the pits in this area. We built new pithead baths at Ellistown, at Snibston – the baths on the opposite side of the road from the pit – and at Measham, which was the last pit round here to have baths.

'Then I went in the army, I joined the Ordnance. I did three years in the army because you were called up for two years National Service anyway and I thought, well, it's only another twelve months and it was extra money. I enjoyed the sport, and I couldn't go wrong. I thoroughly enjoyed it. I was courting the missus at the time, but she wouldn't come out to Germany, so I come out the army.

'When I came home I went and worked back at Orton's again. I was about 21 at the time, and about March time, I was playing football for Ellistown. The pit engineer says to me, "You're a plumber, ain't yer? Do you want a job?" I says, "I've got a job." He says, "What's the wages like?" I told him, so he says, "I can put you on less hours and more money, how's that?" It didn't sound too bad, worth a visit, so he says, "Well, come and have a look."

'I went up and looked around. I knew the baths anyway, because I'd worked on them, and the canteen. I knew them like the back of my hand really. I went to see the manager, Mr Johnson, and the first thing he said was, "Can you put glass in?" I said, "I can." "Start on Monday." I said, "I can't start on Monday, I've got to give seven days' notice where I'm working," and he says, "Start a week on Monday then." He said, "Are you fit?" I said, "Yes, I've just come out the army A1." He said, "You'll fly through your medical then." I put my notice in at the firm. They were a bit upset really, but they didn't offer me any more money.'

Frank Gregory was even older before he went to work at a colliery. He told me, 'I were 25 when I went down the pit and I started in 1957. When I first left school, I was an apprentice mechanic at Forest Road Garage in Coalville. Then I went in the army to do my National Service. When I came out, I went back to the garage but we couldn't manage, because we'd got a first baby on the way. I wanted a bit more money. My father-in-law had worked all his life at Desford Colliery and he said he could get me a job there as a fitter. After I'd done my apprenticeship I went for a year or two to Leicester to the corporation buses. and they set me on as a full mechanic and, after a three months' probation to see if I was all right, I got the top money, which was £8 16s a week. However, it still wasn't enough; so my father-in-law says, "You've got an appointment with Albert King, the engineer at Desford Colliery." So I went and seen him and he listened to me and he says, "Oh, you seem just the job for the fitters here."'

Desford Colliery. (Chris Matchett)

Barrie Hall had three months between leaving school and going to the pit, because his mum was opposed to the idea. 'I started down the pit on 21st April 1952, and we had to do training at Whitwick pit at that time. All my mates from school went down Snibston pit – Snibby – but it took me about three months to talk my mum into letting me go down. So I went to Harlow's for a bit, making boxes, and I was 16 before I went to Snibby to be with my mates. My dad was a coal merchant, and his dad was a coal merchant on the horse and cart, but I wanted to join my mates down the pit. I did four months' training at Whitwick, and then I had a go at all the jobs on the bank, underground, all the lot: then you decide what you wanted to do. I only worked down Snibston, then towards the end, South Leicester.'

Although Barrie went – almost – straight down the pit, he came out and did different jobs before finally returning to mining. 'I went in the army for three years, then when I came out I went lorry driving with different firms until 1974. Lorry driving in them days was all hard work and long hours, and I used to go by Snibby in the mornings and see them all waiting to go down the pit. And by the time I was coming back from lorry driving, the night shift was going down. It was the only place I ever went back twice. Until then, I don't think I ever worked anywhere for more than two years at a time, because once you've been established somewhere, if there were any rough jobs they used to put it on you. You could always get a job then, driving. You could finish on Friday and start a new job on Monday.

'I also delivered concessionary coal [free coal for mine employees] for about two years in Leicester, from the South Leicester depot, but working out of what was then called Leicester Wharf. That was also very hard work, as the coal was on the ground. The mate held the bag on the side of the lorry, and I used to shovel into the bag. We'd do ten to twelve tons per day. A lot of deliveries were to flats. I would try to get the coal bag as high as possible on my shoulder but, by the time I had gone up two or

New Lount colliery, 1960. (Michael Conibear)

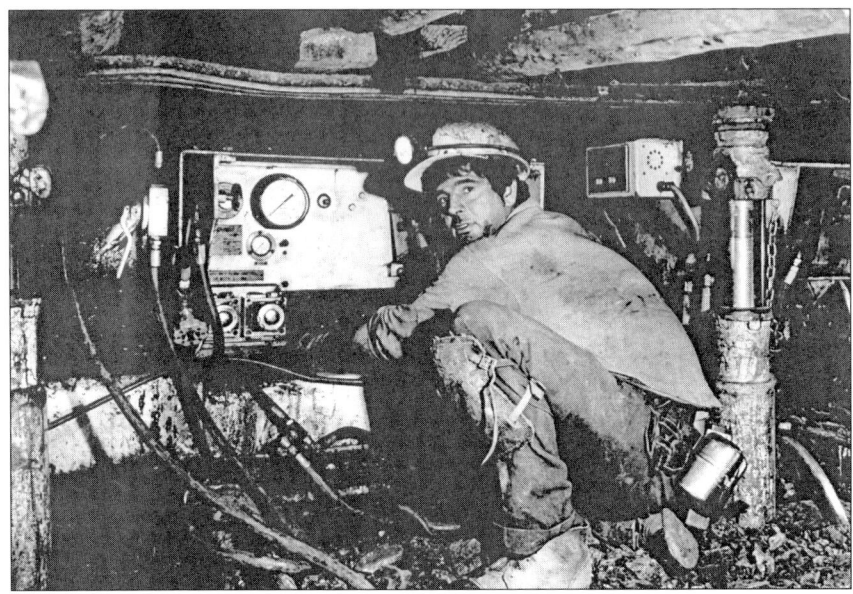

Bernard Hicken with an inseam header at Whitwick.
(Whitwick Historical Group)

three flights of stairs, it was nearly down to my backside. After the 1974 miners' strike, I saw a friend of mine in Ashby, who was an electrician working down Snibby Colliery. I said I wasn't happy driving, and he says, "You want to come back." And it was one of the best things I did. Although it's hard work, and people say, "Why did you go down there again?", you had the comradeship and all that.'

Derrick Holmes came from a mining tradition, but his parents wanted him to go into a different job from his dad. 'I first went down the pit in 1945. I didn't go straight from school, because when I were about twelve, all the lads had got paper rounds. It were like gold dust to get a paper round. My mother happened to be in the Co-op and she said, "Our Derrick's after a paper round but he can't get one, because there's a waiting list." And the Co-op manager, a chap named Frank Manning, says, "I want a lad here three afternoons a week after school." So I started there, and he gave me ten shillings a week. There was only one old man and the manager, and all the others were gels, because the chaps were away at the war, you see. And I used to go and help him weigh up the rations.

'I had to stay at school until the Easter, because my birthday was in January. All my mates had left at the Christmas, but I was fourteen days out. When the day came for me to leave school, he says, "Well, you can have a regular job here. You'll have to go down Coalville and see the area manager." His name was Eggington and his offices were at Market Square. Anyway, I went and I got the job. How much do you think I got a week? Only nine shillings and sixpence, and I'd been getting ten bob for working just three nights a week after school!

'Anyway, I was a full timer at the Co-op. I tell you what, my mam were that pleased I'd got the job there, you'd have thought I were going to be prime minister. Of course, you had to work Saturday afternoons, and of course everywhere you went, you went on your bike. If you went to Coalville to the pictures, you went on your bike. There was no other transport. I stuck it less than six months, and then I went to the pit. My dad had said, "You're not going to the pit. I don't care what, you are not going down the pit." Anyway, I went into the manager's office, about 30 yards from where I lived at Bagworth. There was a row of detached cottages, all the way down, right down to the pit, and the last one was used for the manager's office. The manager had the front room and the under managers had the back. It was now September, so I'd still be fourteen.

'I went in the office, and asked the manager for a job. He said, "Does your father work here?" I said "Yes." "What's his name?" I said, "Jack Holmes." "Does he know that you're coming to the pit?" I said, "Oh yes." He said, "You can start on Monday on the bank. Ask for a Mr Moses Statham. He's the manager on the bank, and he'll tell you where you've got to work." I thanked him, but, when I came out the office, who should be going in but my Uncle Arthur. He says, "Where you been?" I says, "To get a job," and he says, "Your dad'll kill you." Anyway he didn't, and I worked down the pit – Bagworth, Merry Lees and Desford – for the next 39 years.'

Alan Ratcliffe, the son of a miner, had a grammar school education but still looked to the colliery when it was time to find a job. 'I went to Ashby Grammar School, started in 1939, left at sixteen. I had no idea what I wanted to do, so my dad took me along to see the company secretary at Whitwick Colliery – big front room, Sunday afternoon, best suit on. That was it. "Start a week on Monday in the wages office and, when you've been there a while, I'll put you through some accountancy exams." He never did! But I went to the local tech and took Pitman's shorthand, and then an elementary course on accounting, with day journals and ledgers and debiting. I couldn't get the hang of that at all. I took the RSA exam at the end of the year but didn't pass it. So I laboured on until I was eighteen, and got called up. I had the option at that time of either going in the pits or going in the forces, and I opted for the pit. It was the best thing I ever did.'

Alan Pearson had hopes of becoming a professional footballer but an accident put paid to that career option. 'I left Castle Rock School when I was fifteen. At the time I was a fairly good footballer – Notts County had had a look at me, and I'd got a trial at Leicester City. On the day, I couldn't go, and the lad that took my place – Steve Whitworth – went on to play for Leicester and England: I suppose it's a case of being in the right place at the right time. I'd got a knock on my knee and couldn't go, so Steve took my place. I was first choice and he was second. So you never know.'

Alan was another of those lads whose parents were against him working down the mine. 'I actually went and got my engineering apprenticeship at Whitwick pit, straight from leaving school. I went home and told my mum and dad, and my dad said, "Well, you can go and get one somewhere else now, my lad, because you're not going down the pit." He wouldn't let me go down, so I actually went to Burgesses at Coalville, up Belvoir Road, elastic web manufacturers. I went and got an apprenticeship there. My dad never did go down the pit. My granddad was killed in 1941 at Whitwick when there was a collapse and two of them died, and that put him off going down.

'So I waited until after I was married, and obviously had moved away from Mum and Dad, and that's when I went to Snibston. I'd be 28. I says to John Hollick, "I've come for a job, John," because I knew him and he knew me, and he knew my dad. He said, "Is your dad all right about this now?" and I said, "Well, not really John, but I'm old enough to know my own mind." '

Mick Richmond, who was born at Ibstock, was another with a mining ancestry who went into a number of jobs before going to the pit. 'My granddad was a miner at the old Ibstock Colliery – I can remember him coming home from work all blacked up. I was just enthralled by his appearance. I couldn't take my eyes off him, because he was a huge man and he was just jet black. It was amazing. My grandma, down South Road in Ibstock, that's where I was born, she had this big tin tub – and the water had to be red hot. And he'd get straight in. It were red hot. I couldn't believe it. He was like superman to me. I used to just watch him, bathing. He was a miner definitely.

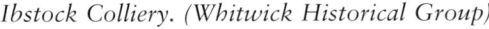

Ibstock Colliery. (Whitwick Historical Group)

'When I was two, we went up to Lancaster. My dad followed the work up there. Then we came back down here to Ibstock on my sixteenth birthday. Leaving Lancaster broke my heart because all my mates were up there. I got a transfer to Halfords in Leicester, the shop in Charles Street, which is now gone. Then I got transferred to the Hinckley branch. When my dad was working at the carpet shop in Coalville, he told his boss, "My son can sell fridges to Eskimos." And he said, "Really?" I got an interview and I got a job. Incredibly, stacks more money than I was getting at Halfords. And I was near home. I joined a group, a rock group called the Merlons, and I did okay with that. We went all over the place. We had an agent in Shepshed, called Danvers, and he got us supporting all these top line acts, including Johnny Kidd, Eden Kane and Peter Jay. It was just unbelievable.

'Then I went to the brickyard for more money and less hours, because I had to work Saturdays in the shop and I couldn't get time off to play in the group. And people kept saying, "You've got to go down the pit, the money's good." And once PLA came in – the Power Loading Agreement – the money was just fantastic. And Bagworth was *the* pit. If you got a job there, you'd be okay. And I got a job and did all my training. I always wanted to go on the face. I loved it. I loved the pit. It was 1977, so I'd be 30 then. I lied at my medical, because one of the things was I'd had rheumatic fever as a child. So I lied about that, because if I'd told them I wouldn't have got the job. Eventually they asked me if I wanted to learn to drive the coal-cutters, the shearers. And I loved it. I was as happy then, happy as a potato down there. It was me. Everything about it, I loved.'

Chapter 3

Rogue Ponies and Ravenous Mice

A **picture called *Pals*, a photograph** of a young collier and his pit pony, hangs on the wall in many an ex-miner's home. There is both comradeship and affection between the two subjects. Some miners – from different areas of the country – claim that the photograph was taken in their own pit, but the truth is that it was taken in 1959 in the main trunk road of Whitwick no. 6 pit by National Coal Board photographer, Donald Ottey. The young collier was 18-year-old Michael Gould from Thringstone. The pony was called Neal, and was named after a Coalville hero: Flight Sergeant Eric Neal, the first local man to win the Distinguished Flying Medal.

It might be thought that pit ponies disappeared from the mines a long time ago, but Frank Gregory, who worked at Desford Colliery as a fitter for 26 years, told me, 'We used to have over 20 pit ponies at Desford when I started in 1957, and when I finished

Pit ponies having their annual sight of grass.
(Whitwick Historical Group)

in 1983, the year before Desford pit closed, there were still ponies then. They gradually petered off from 20 down to perhaps six, towards the end. By then, they had them just for certain places where they couldn't work machines, where the roof had come down low. They had young lads looking after them. They were looked after, you know, well looked after. They'd got their own stables underground. This was after 1947, when the pits were nationalised, before that the conditions were bad for the ponies and for the men. Every year they used to bring the ponies out, heavily blinkered because of the sunlight, coming out for a fortnight's holiday. They put them in the pit fields, grazing there for a fortnight.'

This annual treat for the ponies was also mentioned by Barrie Hall, who worked at Snibston pit for three years in the 1950s, and then again in the 1970s and 1980s. He told me, 'There were pit ponies when I first went down and they used to bring them out in August when we had a holiday. They put them in Pinders Field at the side of Snibby, and they weren't too keen to go back down. It was quite hard work for them underground, but they were well looked after – there was a stable in the pit bottom. The chap who looked after them was called the hostler.'

Derrick Holmes began his mining career at Bagworth Colliery in 1945, two years before nationalisation. He said that, as a boy, 'You never used to see a pit pony out of the pit. Only when I was a child, they did used to bring them out and show them at Bosworth Show. Then after nationalisation, they got a week's holiday, you see, in August. Until then the holiday was only two or three days.' Derrick thought that the ponies seemed to change their personalities when they were on the surface. 'We knew them all, you know. I'll tell you what, there was one particular one, he knew every word you said, but when he got in that field he wouldn't take no notice of you.'

Derrick also mentioned one particular phenomenon that always occurred when the ponies first went back underground after their time in the sun. 'Normally, when they were down the pit, they only ate chaff. But when they did go back down after their holiday

Ted Kerry, the area hostler for Snibston, Whitwick and South Pits.
(Chris Matchett)

in the fields, all their droppings would be green, from eating the grass up on the surface.'

Despite the sentimental scene captured in Denis Ottey's photograph, not all the ponies – or their handlers – were well behaved. Barrie Hall continued, 'There were some buggers. There was one bloke and he'd just let his horse run and it'd run out. You'd see a cloud of dust coming down, then the horse would come past and the bloke'd come by about five minutes after.'

Frank Gregory knew even more about the bad habits of some of the pit ponies. 'We'd got a rogue pony called Tango, down Desford. Most of them were all right but this particular one, he wasn't very big – he was one of the smallest there – but he was notorious. If you didn't know about him, you'd be walking along the roadway with your tools, whistling away there, on your way underground. You'd meet the pony coming back and you'd get in the side to let him go by, and he'd turn – it didn't matter what he were pulling – and if he were in a bad mood, he'd bite you, give you a nip.

'The pit men used to give him sweets to try and keep him happy. He chased me once, I'm not kidding you. He weren't tethered, you see. I dropped my tools and I had to jump on the belt and off the other side to get out of his road. It frit me that much, because the least I'd have got was a good nip.

'There were other ponies, they'd look for food in the gate road. If you'd hung your coat up on the side with your snap bag in, if it weren't in a tin, just in a paper bag in your pocket like I used to take mine for a start, the ponies would go in your pocket and eat it. From your jacket. Eat it paper and all. You know, they'd munch it. The pit men used to like an onion, and they'd eat it raw with a bit of cheese, but the ponies didn't like the onions. They'd go for the good things. It were usually visitors who got caught out, because we got to know them.

'I'll tell you what, on occasions, something would frighten Tango. They might be loading or unloading stuff, and if they weren't aware and they hadn't got a scotch in the wheel – you used to put a scotch in the spokes to brake it – if they didn't do that, he'd break away and set off. He'd been known to be found in the pit bottom, at the bottom of the shaft. He'd galloped all that way, in the darkness, pulling a load.'

It wasn't just the ponies who could break the rules. Alan Ratcliffe, who later qualified as a mining surveyor, began his working life in 1939 as a wages clerk for three days of the week, and the rest of the week he worked underground as a linesman, painting white lines on the roof 'to keep the roads straight'.

Alan admits that once, in his younger days, he did a somewhat dangerous thing. 'Low roads used to be a pain for me, you know, hundreds of yards to go at three foot high. And I was six foot two in those days. It was a real thing on my back. And in some of the roadways – and Whitwick was always a wet pit – the floor was just sludge. The return roadways, where the air was coming out, were the supply roads and they used to take supplies in there with a pony and tub in my days. Two of us once rode out in a tub. The

The blacksmith at Snibston pit. (Michael Conibear)

driver sat on the "limmers" – the shafts to the pony. We had to keep our heads down because the roof was so low. Once it got going, that pony really flew – he knew the road. I found it very frightening and I never did it again.'

Nailstone colliers: ? Tomlinson, Jack Allen and Isaac Storer. (Chris Matchett)

Alan also showed me one of the most fascinating documents I have seen. This was the hostler's diary from 1890, containing the weekly reports on the condition of the ponies. Entries include the following for 3 December: 'I have examined all the ponies in the pit and find them all in working condition except Little Bit, cut on the back, and Tinker, cut under the fore leg.' The entry for Friday 28 December reads: 'I have examined all the ponies in the pit and find them all in working condition except Beauty, cut under the eye.' All the entries are signed by the hostler, Eli Hallchurch.

Ponies were not the only livestock to be encountered down the pit. Frank Gregory told me, 'You'd got to watch out for mice, there were plenty of mice down there at Desford. There was stone dust, put down to stop explosions because stone dust will quench the flames. Well, if you looked down the side in the white stone dust, you could see mouse paw marks and little trails made by the mouse tails. Those mice would go in your snap tin if you hadn't got it properly shut.'

It was no better at Snibston. Barrie Hall recalled, 'We saw some mice. You'd got wires for the man riders, and some days there was some clever dick would hang probably four or five mice up by their tails, for when you got there the next day. However, there was some big hefty blokes down the pit who were frightened of nobody but who were really scared of mice. I've still got my tin upstairs that I used to keep my snap in, because in them days the mice would climb up and get in your bag to try to get to your sandwiches.'

The mice were bad enough during the normal weeks, but when it came to the August two-week shutdown, they used to get really hungry. Derrick Holmes said, 'There was always mice around. I had an experience with mice at Bagworth one August Bank holiday – by then we'd started having a fortnight off instead of a week. Now I was going to Blackpool in September, so I worked the August fortnight. We were blocking a road off. We'd got two chaps packing, and we were ripping into a wooden "danny", a wooden tub, with one end out, and we were dropping the stone down. Now there was nobody down the pit, so the mice were ravenous. And I'm not kidding you, they were running everywhere.

'And this chap named Keir Burton said, "Do you know how to catch em down here?" I said, "Aye, with a mousetrap." He said, "No, I'll show you how. Leave me your water bottle today, and I'll show you how to catch 'em." He put two little bits of bread in this here Tizer bottle, just reared it up, and he left a little drop of water in the bottom. And the next day when we went, it was absolutely full of mice. You'd got to see it to believe it. You see, they'd had no food or water. That were Keir Burton, he was a character. A very popular chap and a good practical miner.'

Malcolm Tudor, an electrician at South Leicester Colliery recalled, 'I was often down the pit during the fortnight shutdown, when the only people down the mine were the electricians and the pump men. I've sat eating my snap with a semi-circle of mice watching me. I've seen half a bucketful of mice caught with one trap. We'd use the same piece of cheese or bait, set the trap, walk round the corner and it'd go snap. Reset it, and snap, another mouse. The mice used to come out from the stables, and they'd get there in the horses' straw.'

It might be supposed that no other animals were to be seen down a coalmine, but the existence of the mice did make one other animal necessary. Derek Howe recalled with a laugh, 'Once I had a bit of a scare. I hadn't been on the coalface long at Snibston, and I was working behind the stables when we'd got ponies. I was looking down and I saw something shining. Two eyes! They kept glaring and it frightened me. And do you know what it was? It was the cat from the stables. And because it was dark, I could only see its eyes. It was really frightening to see these eyes. I thought, What the heck is that? It had come up from the stables, the old cat had.'

Finally, some more cat-related memories from Malcolm Tudor. 'We had two cats down the mine, one in the top pit and one in the bottom pit. I believe one was a male cat and one was a female, but they never met each other. They lived separated. And Min, the female, I've seen her right out in the district in the pitch dark. I don't know if you've experienced total darkness, because even on a dark night on the surface there's a bit of light from stars and so on. But underground, if you put your lamp out, it's totally black. You cannot see something one inch from your nose. But it didn't worry Min. I've even seen her right down on the coal face. She'd actually ride on the conveyer. You know how a cat's eyes glow. You'd be walking into a district, and you'd see these two shining things going up and down. The conveyer goes over rollers, so as the cat travelled along, her eyes could be seen going up and down.

'She did have one nasty habit. The men working in the pit bottom, they'd dug a pit bottom cabin out. It was warmer in there and they'd go in and have their snap. Min used to catch a mouse, and go into the cabin and have her snap at the same time as the men. To eat your sandwiches while she was crunching a mouse up was not very nice.'

Chapter 4

Top and Bottom

Many of the miners I chatted with liked to talk about their jobs down the mine. However, some started their mining career on the surface, working on the screens, where coal was sorted and graded.

One of these was Alan Pearson, who told me, 'I got on really well up there, great bunch of blokes. And it was interesting, because a lot of them had been down the pit. I mean, the foreman, Cyril, he was in a cage that crashed, and he'd got terribly ulcerated legs. But he always used to go to work, right till he retired. We worked Saturdays and Sundays, because during the week they always used to be running, so Saturdays and Sundays we could do maintenance. And believe me, there was an awful lot of maintenance to do.

'I'm glad that I went on the screens because you saw mining from both sides. You got to know the engine drivers and people like that. The last year before we shut Snibby down, I mean we were literally working eighteen hours a day. It nearly killed us. We were there at five o'clock in the morning, because they'd got rid of all the men at Snibby, but we'd still got Whitwick and South Leicester coal coming through.

'I was on the screens a number of years. Working on the screens was fascinating because the stuff they washed the coal in – magnetite – was five times heavier than water. People think a bag of concrete is heavy, but a bag of magnetite was much heavier, believe me. We used tons of the stuff a week, tons of the stuff. The coal used to go through these bars and because of the magnetite, it would float. It then went out over the screens and was sized. All the rubbish and all the stone dropped to the bottom, and that went out on another conveyer.'

One of the first underground jobs that many miners were put on was working in the pit bottom, the area at the bottom of the shaft, loading the tubs full of coal into the cage.

Barrie Hall recalled, 'My first job was in the pit bottom. In them days, of course, the coal came out in tubs. Empty tubs came down,

Tub of coal being taken out of the cage at Whitwick Colliery.
(Whitwick Historical Group)

Screens at Whitwick. (Whitwick Historical Group)

and full ones went up. My first job was uncoupling the full tubs. They came down the road and I used to have to wait while they came round the bend, then just get in between them. It all depended how fast they let them come down to you. They went down to the onsetter and he let them go into the cage. Then on the other side of the cage was where the empties came down.

'And if you did anything you shouldn't do, the deputy would say, "Right you're on the run-round tomorrow." The run-round was a job where, when you were turning coal, the cage goes up and down without a stop. The empties come off, and you had to push them so far till there was a bit of an incline and they could go on their own. And by the time you'd done that, and then got back to the cage, it were there again. It was non-stop, there was no chance of having five minutes until quarter-to-eleven, when it was snap time. It'd stop for fifteen minutes, then off it went again, so it was a little bit of a punishment.

'On the other side of the road where the full tubs were coming down, the empties had to go into the inbye, and that was called "clipping on". You had like a clip on an endless rope, you used to put that on, and as I took my couplers out, the bloke on the other side of me, he'd be coupling the empties. It was just like a factory, that pit bottom work. I didn't do that long, because everybody wanted to get on the face, and do face training. I did that for a while, then I had an accident and I was out. Mothers in them days had a bit more say than nowadays, and if your mother said, "You're coming out," you did.'

Derrick Holmes moved from the screens on the surface at Bagworth Colliery into the pit bottom. 'I started on the screens, picking off. I did that about six months, perhaps a bit more, and then I went down the pit. My first job down the pit was scotching up, in the pit bottom. We used to stop the tubs with scotches – brakes. The onsetter put the tubs into the cage, and we let two tubs at a time go to him. It was a nice steady gradient. I would have to hold the tubs back while he let two go in. They took some holding on wet days. They used to make their own scotches at Bagworth. They cut a wooden bar and two niches in it, slots about the width of the rail, so it was like a little cricket bat, with a handle on it. You used it as a brake by putting it on the rail in front of the tub wheel. We used to do about 500 tubs a day.'

Long before Alan Ratcliffe qualified as a chartered mining surveyor, he had a double job at Whitwick pit. For part of the week he was working on the timesheets for the wages office, and the rest of the week he was a linesman. 'We had to do the time books Fridays, Saturdays, sometimes Sunday. We'd do them down the pit. The deputies would have a time book and they'd fill the men's hours in every day. We'd go down on Friday and copy these into our big ledger. Make them up, so they'd be ready for the wages people on Monday, and they'd copy them yet again. Very labour intensive. After Monday, we would be doing some stats for the manager, output – output-per-man-shift – and various figures.

'Then on Tuesday, Wednesday, Thursday, I'd be down the pit painting lines on the roof, to keep the roads straight. We were linesmen. We'd go down one gate road, and there'd be five or six yards needing lining. There'd be some old lining but it'd be a mark here and then another mark there. We used to average, and hang a clothesline up so it went through the majority of these marks. Then we'd go along with a safety lamp hanging on the line, and the lamp would throw a shadow on the roof. And when you'd got your shadow going through most of these marks, you'd then go forward and paint the new line. You'd paint the shadow. One-inch brushes and a tin of whitewash – splosh, it was known as. It was okay if you could reach the roof, but sometimes you were clambering on anything to get up and reach it. And when we'd done that we'd move on to another district and so on.

'Then if there was any surveying work to be done, we'd be helping. Whitwick hadn't got a proper surveyor and the under-manager did the work, but a surveyor named Horace Shackleton would come out every three months to do the statutory quarterly survey and we would be the people who would make up his team. I'd be the back man holding the lamp at the back station, while the surveyor was up in front. He'd give you a certain signal, moving the lamp in a circle, and when you got that signal, you were clear and you could come up. I'd be eighteen, then.

'People sometimes used to say, "Oh, you weren't filling coal, you weren't a real miner, you were only down the pit," but it didn't make any difference whether you were filling coal or not. We were all miners. A miner is a bloke who works down the pit, isn't he, whatever his job? A falling lump of coal didn't differentiate between a bloke filling coal and a chap with a surveyor's dial. If it came, it came, didn't it? In fact, one linesman at Snibby got killed when a piece of coal trapped his head against a prop.'

Working in the pit bottom was all very well, but what many young miners really wanted was to get to the coalface, to have a 'ratch' of their own. A ratch was a length of coalface, about nine yards long, marked by the deputy with a chalk line on the roof. Each ratcher would shovel the coal from his nine yards onto the conveyer belt. Young miners regarded the job of ratching as the pinnacle of pitwork. If you had your own ratch you were a real miner.

When Alan Ratcliffe was working as a linesman, he liked to get down to the men on the coalface. 'When we'd done our lining, we didn't go right back, we'd go down the coalface. The men would be turning coal, there'd be ten men on, each doing a ratch – a length of coal. And as you went down, if you were going in the direction of the belt, then you'd ride the belt. You'd keep saying, "Hold off! Hold off!" else you'd get a shovelful of coal on you, because they didn't want to stop while you went past. But if you were going against the belt, well, you'd have to crawl up the face. They'd say, "Come on, chaps. Stop and give us a chuck on, get this shovel." And at eighteen, it was an adventure. I used to stop with them and do some shovelling.'

Derrick Holmes told me of a trick that the ratchers at Bagworth would do, so that they didn't fall behind, even when the conveyer belt stopped running. 'When the face conveyors stopped, the ratchers kept sneaking a shovel full or two onto the belt. This was known as bauming, but it could cause the belt to stall.'

Joe White, who grew up in Whitwick, wanted to work on the

New Lount Colliery. (Whitwick Historical Group)

coalface. 'My first pit was South Leicester in 1976. I started on haulage systems, then I was quite fortunate really, and that was down to the fact that South was an older man's pit. We were young teenagers, but there was a lot of older men coming up to retirement. We got the chance then to move into their shoes – their pit boots – to develop into face training.

'After I'd only been there twelve months, I went on face training, which at that time was 130 days. At the age of nineteen I went on to face development work. It was all automated production work, machines and power supports. But within that, the skills of the older miners still left at South Leicester were very valued. Because, even with what machines we had then, all of them old skills was invaluable. They really was, because you can't beat a pick and shovel, and sometimes a bit of a cleat, you know: "Knock a bit of wood in it, kid." Because you're working with Mother Nature, who has some cruel ways to catch you out. Those old skills carried me on, well, all of us, all those young miners, to thankfully a safe career.'

Mick Richmond was pleased to get a job at Bagworth, especially when he got a chance to drive the coal cutting machines on the coalface. 'Bagworth was a modern pit, you see. The seams at Bagworth were 12 ft, they were huge. I could stand up in the roof supports, me with a helmet on, and drive a machine. And that's how high the walls were. It was amazing, it really was. They called the seam "the five foot and splent". What it was, there was five foot of coal, then there was about a foot of what they called a muck band, like clay and whatever, then there was another five foot of coal. We had what they called a ranging drum shearer. Say if you were shearing that way, the front disc would be up in the air, cutting the top coal, and the trailing disc would be cutting the bottom coal. I was on that face about five years and everything about it I loved. It was me.'

Whitwick Colliery. (Whitwick Historical Group)

There were some specialised jobs at the pits. 'Ozzy' Osborne started at Ellistown pit as a plumber. 'I had my own little workshop, I was my own boss virtually. I did the maintenance on the baths, the medical centre, and the canteen. Then there was the lamp cabin, the powder magazine. I took over the pipe work and boiler maintenance. They were steam boilers then, and the steam worked the winding engine for the cage. I did that for quite a few years. I was the representative for the fitters department on the safety committee. I was the captain of the fire fighting team, and we won the area competition several times. I was playing football and cricket. I was all right, doing very well. I was on surface rate unless I went down the pit, helping the blacksmith and the shaftsman and cable changing, and things like that.' Ozzy later went on to become a shaftsman himself.

Alan Ratcliffe, who had started as a linesman, decided to train as a surveyor. 'My wife-to-be said to me one Saturday, "You're wasting your time at that job. A grammar school lad with a school certificate, you're wasting your time." She was right and it sort of woke me up. I'm a great believer in woman's intuition and instinct and sensibility. That August there was an advert for apprentice surveyors in the *Coalville Times*. When I went surveying, gave up my timekeeping job, and took this apprenticeship, it meant a drop in my wages. I had been taking home at the age of 22 about £9 a week. And when I got the apprentice job – I was too old to be an apprentice, so they made me an uncertificated surveyor – I dropped down to seven pounds a week. It was nearly a 25% drop. But my word, it's paid off since. It paid off hands down, including my pension.

'I got an interview at Nottingham tech, and they set me on. I was in a class of four, so I was important to them because they couldn't run a class with less. I used to go out and catch the half past seven bus out of Coalville to Ashby, pick up the X99 Ashby to Nottingham, get there just before nine o'clock. Then at night, eight o'clock I caught the bus back to Ashby, then Ashby to Coalville. I walked in home at ten-thirty. So it was a stint. This was every Wednesday, one day a week to start with.

'By that time I'd got married, and it could have caused matrimonial problems because my wife had been at home all day. I'd be fed up with listening and talking all day and I'd want to bury my head in the paper. She'd want to talk. Never was a row, but it was conflicting interests. And so it went on. I got sorted out in three years and took my ticket.

'I was older – I'd be about 27 – and the door was closing on pit surveyor vacancies. Every pit by that time had got a surveyor, and it was only by movement in the ranks that there was a vacancy. I qualified in the January and in the December, Lount and Swadlincote came vacant, because the new area surveyor had set up group surveyors, which we'd never had before. Leicestershire had been a backward area for surveyors, particularly Whitwick. The surveyors from Bagworth and Bretby were given Lount and Swadlincote, so it created two vacancies and I applied for them. I went for the interview and it turned out I'd got Bagworth, which was a lovely little pit. The camaraderie was wonderful. I had thirteen years as surveyor at Bagworth.'

Derrick Holmes, also from Bagworth, was another man who had to make a decision about taking a pay cut in order to go for promotion, but he was less keen than Alan. He'd been very happy working on a ratch next to his dad, and then going cutting on a face, and driving a new road through to Nailstone. He was earning good money going in on Sundays on a different job. 'I was getting £4 a completed yard, and by this time, we'd got four children. This was 1960. I came out the pit one day and the manager, Mr Agar, called me in and he said, "Derrick, I want you to get your deputy's papers." I said I couldn't afford to go as a deputy, because a deputy was only on about £16 a week at that time, and I was on top money. He told me that I would still be paid my heading rate. Anyway, I went and got my deputy's papers, but of course I stayed on my job. I'd got my deputy's papers but I weren't a deputy, I was still a workman.

'Anyway, this went on for about six months, and Mr Agar was moved to Desford Colliery, and a chap named Bird came to Bagworth, Dickey Bird we called him. I'd got to go and see him, and he says, "What do you think you're on, working down there?" I began to say, "Well, Mr Agar said ..." and he chipped in, "Mr Agar's not the manager now. I am." He said, "You'll be deputy on forties face on Monday morning, Derrick Holmes." I said, "Mr Bird, I can't possibly live on that wage." He said, "It's cost the Coal Board a thousand pounds to educate you to be a deputy, so that's what you'll do."

'I came home and my wife Blanche went and got a job, because of the drop in pay, and my mother, who lived up the road, came and looked after the baby. This went on until I was that down that I went and put my notice in. I hadn't got a job, and I'd got a family and I'd got a Coal Board house. I thought, I've dropped myself right in it here. Then I was walking up the road towards the club, and I met this Bagworth under manager named Alf Henson. He called out, "Ay up, Decker. How you going on?" I said, "Not too bad." He says, "What's up wi' yer?" Now this Mr Henson had been the under manager at Bagworth pit, but he'd left and gone to Merry Lees. So I said, "I haven't got a job. I've gave my notice in and I'm out of work." I told him the story and he said, "I want you to go on Monday morning down to Desford pit." This manager had been to Bagworth pit and seen how we were working, because we were doing it a bit special.'

Derrick went to work on development, opening up a new face at Merry Lees, then worked at Desford. When the deputies gained a big increase in pay after the 1970s strike, he became a deputy again, and ended up as an overman: 'That's how I finished my days. When Desford pit finished, by this time we'd got a manager, but no under manager. I was the overman and I had six deputies and 40 workmen on salvaging. I took advice and I retired when Desford Colliery closed in 1984. I could have moved to Bagworth pit, which was still going, but I didn't.'

Chapter 5

Smells and Bells

The mines all had underground systems for the circulation of fresh air and for warning signals. Working in the pit could be a chilly job, because of the inflow of cold air coming down the shaft. The air came in the downshaft, travelled all through the inbye roadways, down the coalface, then along the outbye roadways and out of the mine through the upshaft.

The air coming through the downcast shaft would be cool and fresh, but after it had travelled down the coalface and was on its way out, it would be warm and carrying a number of smells. Some of these might be quite appetising. Mick Richmond remembered one story: 'When I was passed out as a machine driver, that's the first time I met Cliff Jeffrey – they call him Geek. He was a machine driver, a very good machine driver, and when I first saw him on a machine, I was nervous, because he was so good. And he used to wap the 500 horsepower coal cutter up, and here's me going quarter speed, half speed, and he'd be flashing his light. I'd stop my machine and say, "What are you flashing your light at?" He'd say, "We've got to go faster than this." Anyway, we got on great.

Whitwick colliers: Bernard Hicken, Arthur Sykes, Mick Starkey, Ken Bradley, Henry Burford. (Whitwick Historical Group)

'The tiny flecks of coal, they gradually build up over the day's shift, so when the afternoon shift came on, there was like a mound of this coal. Now this coal machine was a huge thing, and when we were on afternoons, he'd say, "Just dig enough out there for a biscuit tin." So I dug out a space on top of the machine, and he got this biscuit tin out of his snap bag. He says, "What about this then, kid?" and he undid this silver foil and there were these hot dogs. Bread rolls with sausage and onions. I says, "What are you going to do?" and he says, "Put the tin on there, on top of the machine." This machine was red hot. He says, "Cover it all up now and about five o'clock, they'll be fantastic." And I said, "I can't believe this is happening."

'And at five o'clock, we stopped the machine and we had us snap. I lifted the biscuit tin up, burnt my fingers because it was red hot. Down the pit, anybody'll tell you, if the air was coming up

'Going Home': Joe Hall, miner at South pit, 1934. (Chris Matchett)

the main gate, it came across the coal face, then back down the tail gate. So the people in the tail gate knew exactly what the people on the face and in the main gate were eating. It was that strong, you see. Pea soup was one. If someone had got pea soup in a flask you could smell it. Or oranges. And Geek said to me, "Listen in a minute to the dacks." That's the tannoys. So we got the tin lid off and we'd got onion fat, and sausage fat, it was all over him. Suddenly a voice comes, "Who's got sausage and onions up there?" I was laughing my head off, honestly. It was the best hot dogs I've ever tasted in my life. That was a happy memory.'

The smell of hot dogs might have been a pleasant one but some aromas were less appetising. Frank Gregory told me, 'I remember one particular trick, it wasn't very nice really, and that involved unsavoury smells suddenly coming down because somebody had been to the toilet in the gate road. If you can imagine, there's the face where the coal's cut, where the miners work, and there's two gate roads, that's the access roads – the supplies and the men go in one and the coal comes out the other one on the conveyer. That's also the air flow ventilation. For some unknown reason, they seemed to like going into the inflow to go to the toilet, where the air was coming in, so we all got the benefit, which I didn't particularly like. In headings, which of course had just one road in because they were developments, they put a fan in and a long nylon tube to feed it. They were very powerful fans to give you plenty of air – if you stopped working you were cold. Well, you have no idea what that's like if somebody goes to the toilet and it comes down the fan. And that's where the old saying comes from, when the so-and-so hits the fan. That is exactly where it came from. That's one of the worst parts of it, but we used to have some fun.'

Alan Ratcliffe confirmed Frank's memories. 'Anybody with an orange, you could smell it hundreds of yards away. You could also smell much worse smells. The place not to go on a Monday morning was down the coal face. Not to be advised. Not when fifteen men had been on the weekend beer, and it all had to be got rid of. They'd probably had brussels for dinner as well. They'd come back from the pub at two o'clock on a Sunday, eat their dinner, go to bed, then out down the pub at six for another round. Some of them were real hard drinkers. You do know, don't you, that down the pit, you'd never pick up a piece of newspaper lying at the side of the road? Because it had been used. I used to carry a bit of paper about in my pocket, because down the pit you'd have to go sometimes, particularly if you'd been down all day. There's no stopping, you've got to go, haven't you? If someone was picking stone out of the gob for packing, they'd often put their hand in something unpleasant.'

Alan also told me about the overhead wires that could be used to signal that something was amiss. 'I know one little incident when I was lining. You walked about the pit on your own, you didn't go in pairs, so it was a bit lonely. I was walking down a haulage road one day, an endless rope haulage, where the empties were going in and the full tubs were being brought out. You had to be careful if you got near one when they were passing, because they could roll

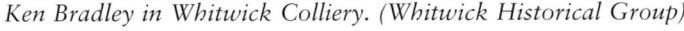

Ken Bradley in Whitwick Colliery. (Whitwick Historical Group)

New Lount, 1957. (Michael Conibear)

you between them if they were too tight. So you'd let them pass each other. You had to keep your wits about you. And overhead would be bell wire, a double wire, two separate bare wires.

'I was going down this road one day, on my own I thought, and I heard a tub coming up the road, clunk, clunk, clunk. And as it came along, I could see the back wheels bouncing. It had got off the track somehow. So I got my penknife out, reached up to the bell wire, pulled them together, put my penknife across, and that rang a bell back in the engine house at the end of the road. It stopped the job. And a voice suddenly said, "Well done, Alan. I can see you're getting your pit legs." That was the overman. I didn't know he was behind me. I knew what to do, but if you were a stranger to it, you wouldn't. Between us, we got the tub back onto the track. But that could have derailed another lot, and they could have had a pile up, and stopped the job.'

Working in the upcast foreshaft area of Asfordby mine. (Chris Matchett)

Those overhead wires could also be used for a bit of mischief. Barrie Hall gave me an example of winding up the deputy at Snibston pit, telling me, 'At Snibby, there was a pit deputy and he used to be in charge from the pit bottom to a place called Top Turn. One day someone would cross the wires and stop the haulage from going. The deputy then would have to go inbye to see what was happening. About half a dozen men worked down there, one clipping on, one coupling up, one uncoupling, one knocking off and so on. What they'd do was cross the wires somewhere where it couldn't be seen, and the deputy would go by and say, "I'd better go and see what's going off." The men would sit in this little hole in the wall and have a little laugh and talk a bit. Then when they thought the deputy had walked nearly to the end, they'd uncross the wires, and the belt would start up again. And the deputy would come walking back with a sweat on and they'd say, "What was up, Bert?" He'd say, "I don't know, it came back on before I got there." We'd all be laughing and thinking, Aye, we know.'

Chapter 6

Greyhounds and the Miners' Welfare

The miners and their families lived, worked and relaxed together in a tightly-knit community. Many of the miners' hobbies involved competition – growing the largest vegetables or breeding the fastest racing pigeons. Another interest was greyhound racing.

Derrick Holmes was born in the mining community of Bagworth, and worked at Bagworth, Merry Lees and Desford pits. Derrick told me, 'My interest in greyhound racing began at the age of about eighteen, when my father and a couple of his mates arranged a trip to Leicester Stadium one Saturday afternoon. From the age of 21 I was a regular on the terraces of the cheap side at the stadium every Thursday and Saturday. Two workmates, Granville Poole and Ron Saddler, were also regulars and we would talk of nothing else as we walked from the pit bottom to the coalface.

Derrick Holmes was a regular on the terraces at Leicester Greyhound Stadium.

'The first dog I owned, I bought from the show jumper Ted Williams, out of his bitch Marlboro Beauty, which had won the television trophy. The sire was Faithful Hope and he still holds the fastest hand-timed Greyhound Derby trophy record. My enthusiasm increased as my pup grew. My two sons, John and Derrick, and I would take it out to the fields. I'd made a lure from an upturned cycle frame and one son would wind the pedals and the other would release the dog.

'My dad bought my next dog, which was a pup from Ireland. From this time on, I started to go to all the local tracks: Coalville, Hinckley, Long Eaton. My boys and my youngest daughter always came with me. Heather was about eleven at the time. She used to parade the dogs and put them in the traps. What a thrill when you had a winner.

'I loved coming home from work and taking my dogs for a good walk. My wife, Blanche, always fed and walked them along with the children while I was at work. I am sure that this hobby helped me while down the mine. The plans made for a future race were never far from my thoughts. Over the years I had ten dogs. One I raced at Leicester and one at Nottingham. The most prestigious trophy I ever won was in a Christmas Marathon over 700 yards at Hinckley. It was the Don Pare trophy. I won that with a dog I bought at Hackney Sales, and we called him Jack Spratt. My son John took him in that night. After he won the race, we were both so full of emotion we could not speak. I must mention the part my wife played in support. She kept the kennels spotless. Happy days.'

Derrick Holmes with his greyhound trophy.

I was surprised to be told that many miners supported foxhunting. I had always regarded hunting as a leisure pursuit solely of the wealthy and well to do, but Alan Pearson, who worked at Snibston Colliery and later became a councillor on North West Leicestershire District Council, informed me that I was wrong. 'Before I left school,' he said, 'my two uncles who were at the pit, their favourite pastime was foxhunting. And I always said when the Tories shut the pits, I said to the local Tory MP at the time, "You'll regret this you know, shutting these pits down." He said, "What do you mean?" and I says, "The biggest allies foxhunting has got are the miners. Once they're not there, hunting won't have any bloody allies." A lot of miners went foxhunting. I can remember walking over fields to watch, bitterly cold days. I tell you what one of our favourite tricks were, digging a swede up, and just cutting a slice off and eating it. It were frozen, you literally had to suck it before you could chew it, because it was that hard.

Ibstock band flyer.

'I'll tell you another thing you may not be aware of, but a lot of the miners used to spend Tuesdays at Melton cattle market. I used to go, myself. My uncle took me round all the markets: Loughborough on a Monday, Melton on a Tuesday, Leicester on a Wednesday.'

Another leisure activity in the mining community was playing in the colliery brass bands. North-west Leicestershire may have lost its collieries but it still has two brass bands. Desford Colliery Band was founded in 1898 and was originally called Ibstock United. By 1912 it had appointed a professional musical director, who travelled all the way from Leicester twice a week by pony and trap. At the time Ibstock had its own colliery, but it closed in 1929. Ibstock United band continued but after the Second World War, it needed to replace its ageing instruments. In 1956, it received financial aid from CISWO – the Coal Industry Welfare Organisation. At that time, and in view of the fact that most of the band members had always been miners, the band's name was changed to the Desford Colliery Band.

In 1970, the band gained promotion to the championship section, and since then has won more than 30 championship titles. Desford is universally acknowledged as one of the top three colliery bands in the UK, along with Grimethorpe Colliery Band from Yorkshire and the Cory Band from South Wales. On Sunday, 28 January 2007, Desford Colliery Band defeated eleven others to take first prize in the Mineworkers' Championship held in Skegness. Internationally, it is in the top 16 brass bands of the

Desford Colliery Band, 2004. (Peter Smith)

world, and Peter Smith, the band's chairman, tells me that it is rated by some organisations as being in the top ten. The band also starred in its own six-part television documentary programme, *The Real Brassed Off*, which told the story of its day-to-day running.

The Snibston Colliery Band, now known as the Snibston and Desford Colliery Brass Band, has recently been promoted from section two to section one of the brass band leagues. The Desford and Snibston colliery bands, along with the academy band – the learners – rehearse in a building known as the Brass House, located in the former Whitwick Colliery wages offices in Coalville. Formerly, they used the Coalville Miners' Welfare at Snibston.

The Miners' Welfare was one the most active centres of the mining community. They were clubs jointly administered by the Coal Board and the National Union of Miners (NUM). When they were first set up they were dry – teetotal – but were later allowed to sell alcohol. They became a place where the miners could go for

Snibston Colliery Silver Prize Band. (Chris Matchett)

a drink after work, or take their family for an evening's entertainment. As well as the bars, they had a large room that could be used for dancing, or for watching a show. The colliery brass band could rehearse there. Coalville had its own Miners' Welfare, situated on land belonging to Snibston Colliery.

Alan Pearson told me that originally the club was in a wooden hut. 'You always went and had a pint with your workmates down there. There were an awful lot of characters down the pit, and it is sad that you don't get that sort now. It was the breed that you'd got down there. It was the friendship. You'd have a good argument, and then all go for a drink down the club, the Little Club, we used to call it, at Snibby. It used to be a wooden building at one stage, then they built the big complex. Again, it was well used. You could take your family down there.'

The Little Club in the wooden hut was replaced by a magnificent building that opened in the summer of 1965. Although it was a wonderful building, Alan Pearson thought it could have been even better. 'When they first said about having a Miners' Welfare they talked about a swimming pool and everything, but in the end they went for what they went for. While it was a good facility, you just think about what it might have been.'

Derek Howe, a miner at Whitwick and an elected official of the NUM, told me, 'I finished up as a trustee on the Miners' Welfare, because the NUM were very involved. We had all the big bands there, including Joe Loss. Joe used to get in touch himself, you know. He'd come to Coalville and practise in the afternoon, with tea laid on for them, and a cheque ready. Joe was the only one who got in touch himself, the other bands used to come from an agency. There were some good bands, too. I remember once, Ivy Benson's band broke down at Watford Gap. One or two of the band, professionals who used to play for her, had turned up separately but they were sat here in the Welfare waiting. So we sent a coach out for Ivy Benson's band that night. We had most of the top bands.

'And the comedians. We used to have those on Sunday nights. Charlie Williams was one, and Jim Bowen, the chap off *Bullseye*. They'd come here, then go on to a nightclub. They used to leave

Coalville Miners' Welfare. (Chris Matchett)

us at about half past nine, quarter to ten, and go to Birmingham, to do another show – £100, that's what they'd charge us, for a Sunday night. And the club would be full.

'We had some very good times, and that's what kept the miners, kept Snibston, very close together. It was always a very close-knit community. It made a big difference, having the Miners' Welfare, such a big difference. We all enjoyed it, we all had us nights. There'd be a rota and we all used to go. I used to have dancing, because me and the wife loved dancing. Especially to Joe Loss. You always had to have somebody from the NUM side in charge. If the secretary wasn't there, it'd be a committee member. You can't run one of those organisations without you do it like that.

'It was a wonderful club. We had football, cricket – we had a very good cricket side there. When Leicestershire County Cricket team used to go out to different areas, to Ashby and to

Whitwick Colliery Football Club, 1948. (Whitwick Historical Group)

Loughborough and what have you, they always played at the Miners' Welfare in Coalville.'

When the pits shut, the local council did try to buy the Miners' Welfare, but they couldn't raise enough money. Derek Howe explained, 'I remember, as a councillor, four of us went along with officers to meet Jack Jones [secretary of the Leicestershire County NUM], who was still a trustee, and the directors, the financial director and the director, to see about taking over the Miners' Welfare. Jack says, "How much money have you got, Derek?" and I said, "£50,000." He says, "That's no good. You might as well go home as have a meeting." I says, "You're joking." And I went to the officer who was with us, and said, "That's it." So we didn't have the meeting, we come back home, else otherwise it might still be open now – £50,000 wasn't a lot but it was as much as we could afford. Jack needed in the hundreds of thousands, because he wanted to build a similar place over at the new Asfordby pit.'

The Coalville Miners' Welfare had closed in 1988, but in 1991 the building was destroyed in a fire, and houses have been built on the site. Derek Howe has a vivid memory about a coincidence connecting both the Welfare and the houses that replaced it. 'Strange things do happen. As the Chairman of Housing and also as Chairman of East Midlands Housing Association, I was requested to go and turn the first sod at the Miners' Welfare site for the new houses. And that very morning, I was tapped on the shoulder and told Jack Jones had died last night. That was such a big coincidence. It made me think a lot. It brought memories back. Coalville really lost something there. It would be good for the youngsters now.'

Living in a mining community had many positive aspects – the friendships, the comradeship, the brass bands and the Welfare – but it could have its problems. One of these was subsidence. If there were mines under your village or town, the houses could suffer.

Frank Gregory, from Bagworth, described the devastation that his village suffered. 'Let me tell you about subsidence. It wasn't

An aerial view of Coalville Miners' Welfare. (Whitwick Historical Group)

much of a problem until 1947, but then there was a change in the type of mining that the Coal Board wanted to do. The machines got bigger and more powerful, they were ripping more and more coal off. Getting bigger profits, oh that was all right. But they found difficulty because they could only work one shift turning coal, because all the preparation work took two shifts.

'Back to subsidence. Round here the ground dropped about 27 ft. That might sound excessive, but when you think that six seams had been extracted, one below the other, you can soon add them up. The surveyor told me it was a 27 ft vertical drop – across the fields mainly, but there was subsidence in all the old houses. Three-quarters of the houses in Bagworth were wiped out. I don't think I'm exaggerating. There was an exodus of people to different villages.

'What made this terrific change was that the Coal Board said we're not having props up any more in packs, because when they extracted coal they usually put packs up every few yards to hold the roof up, and then they moved on, taking slices. Right. They said we're not putting packs up any more, we needn't pay anybody for doing that, we'll just let it drop. That's called crash packing. They stuck to that until the pits closed, because they were making hand over fist with money. They just kept slicing coal off, pushing over, and within 12 ft of the face we were letting it drop. You had to keep away from the gob, the waste, because suddenly it'd drop with a terrific rush and there'd be a waft of air that'd hit you. The dust it'd stir up! It frightened visitors to death when it happened, you know. Anyway, that was profitable. What happened, the houses started dropping like flies, one after the other.

'The church went. Bagworth church. I was married there, all the kids christened there. It collapsed through subsidence. That hurt us. The tower was Anglo-Saxon and the main part of the church Norman. We weren't very pleased when they said it'd take too much money to do up, it'd be non-viable, you'd be better to let us give you £75,000 and you can have a prefabricated one, and that's what they did. Concrete sections, looked terrible, and 40 years later an engineer's report said that it was unsafe and had got to come down. The only way we can afford to have it taken down is to sell the land. It lasted just 40 years, and the old church had stood over 900 years, made of stone 8 ft thick. The bulldozers had a job to knock it down, they kept bouncing off it.

'And we lost the school through subsidence. That was a big controversy. They built a new one at Nailstone and the kids all had to go there. We wanted one built here in Bagworth, because we knew at some stage the village was going to expand again. The women got onto the council and the county council, but it didn't make a blind bit of difference. This community really suffered through the subsidence.'

Chapter 7

'There Was This Character...'

'**There was a host of characters,**' recalled Joe White. 'Things were serious, yes, but everything was laughable. You just enjoyed your day. Okay, sometimes you'd go to work a bit pissed off because your girlfriend's fell out with you or something else, but with those characters you worked with, and older characters, you just forgot about it. It was just great to go to work. The humour, the comradeship, fantastic. It was a way of life.'

Many Leicestershire miners remember the interesting – eccentric, in some cases – characters they worked with down the pit. Malcolm Tudor, who worked down South Leicester Colliery, told me, 'I actually grew up in a terraced house at 4 Victoria Road near Ellistown Colliery. One of the neighbours was Bill Atkins. You know how you sometimes have a paper that needs to be signed by a responsible person. Well, Bill was in charge on the surface, what we called the pit bank. Somebody asked him to sign a paper for them, so he did. And under his name he added – *Bank Manager!*'

Alan Pearson told me about some of his favourite characters at Snibston. 'I'll tell you one character at the pit, Wally Castle. You know the trowin – the plastic pipes that used to circulate the air round the pit. We cut a piece off, put a bit of sackcloth at the front and painted a face on it, and chucked a cap lamp at the back. It looked like a giant caterpillar, going up the belt. Wally saw it and he run out the pit! It took three days to get him back down, frightened him to death.

'Wally, if he was 5 ft, that's as much as he was. But he was strong – he was a bouncer at the Grand in Coalville. Wally, he lived with his mam on Margaret Street, never got married. We had a laugh with him, because he'd always go to Blackpool for his holidays. And when he went out, believe me, he was immaculate. When he was a bouncer he always used to have a dickey-bow, black suit and everything – the works. His shoes used to gleam. He did a lot of ballroom dancing. He were a character.

South pit men: Alan Reid, Len Bramley, Don Kendrick,
Dennis Gray. (Chris Matchett)

'Another one was Joe Vesty, he worked on the screens. He was a nice bloke, ever such a hard worker. If you were shovelling with Joe, it was a nightmare because he just kept going and going. It could be a pile as high as this bungalow, but Joe wouldn't stop until it was all gone. If you sat down at the end, you'd be sweating, it'd be puthering off you, and he'd say, "Come on lad, we'd better go and find something else to do now." Then there was Steve Ballard, that was on the winder, he drove the drift belt – he was a character in himself. That's what saddens me now. I've worked in offices, and the characters aren't there any more. I went from the pit to East Midlands Housing on a temporary contract and I stayed there twelve and a half years, ended up in development, but the characters are not the same.

'Then you'd got men like the foreman on the screens, Roy Ford. He'd got twin sons, Jim and John. Now, if they sat there you couldn't tell 'em apart, I'm not kidding. Roy did, their dad, but if you wanted them you shouted, "Jim-John!"' because you never knew which one you'd get. It's things like that, that stick in your mind.'

Barrie Hall also had memories of characters from Snibston. 'There was one guy, Joe Wileman, he'd always got a couple of packets of chewing tobacco in his hat. He chewed it and the juice ran down his mouth and all over him. Joe gave me some tobacco when I was doing my training, but I had to spit it straight out. He says, you've got to chew this, it keeps your mouth wet. But I never did. Joe used to come out the pit, and he'd go straight across the road into the Queen's, which was the pub that was there at that time. And he could dash about seven or eight pints down in an hour "to settle the dust". They were hard men. And they could drink.

'I used to love the pit canteen, especially on a Friday. It was full of everybody, not just miners. There were postmen used to go down there. There'd be retired miners, who'd go down for their pension and just to have a chat to their mates. There were even ladies – wives of retired miners or miners' widows – who walked up to Coalville market at that time, and they'd stop in and they'd

Jack Miles in the pit canteen at Whitwick when it first opened.
(Whitwick Historical Group)

have a cup of tea. The canteen on a Friday afternoon was fantastic. I used to really enjoy it. It was a good atmosphere, especially when you'd see all the old miners. I mean, I thought I'd done some hard work but them who were there before me had the really hard work.

'Then there was the borer, who carried the drill – they used to called it the "pig's head". He'd pick the pig's head up and do the drilling for the ratches. Ratching was hand-filling. Each man had a ratch or stall of 8 or 9 yards of the face, and he'd shovel the coal onto the belt. You'd got your own ratch there, and there'd be another one, and another one all the way down the face. The borer's name was Albert Burkin and he used to chew 'bacca, and he'd got 'bacca all over his face, all over his pig's head and everywhere. Good lads, they were, all good lads. The best thing was the atmosphere, the friendship. There were arguments, but it

85

all seemed to come back to normal when you got up on the bank.

'The things they used to try and get away with, to get out of the pit early. As soon as they come out, they'd all got little hiding places for their cigarettes. And if you'd got some it was "Lend us a fag, give us a fag." And a packet could soon go.'

Derrick Holmes added, 'There used to be a character called Horace Clamp, a good old workman. We'd say, "How are you, Horace?" and he'd say, "Ready for the gun".'

Alan Ratcliffe had many recollections about the men he'd worked with. 'When I first went down at Whitwick, lots of men would take snuff. I suffered with my sinuses at that time and I used to take a medicinal snuff called Besorbin. I carried it with me in a little tin. Lionel Ball, he was about my age, and he was the pony driver. I stopped chatting to him one day and he says, "Want a pinch of snuff?" "Yes, I'll have a pinch with you, Lionel," I said. Now this was miners' snuff, the brown stuff. They used to sell it in the canteen. He tipped a heap on the back of my hand, and I sniffed it up one nostril, then up the other. Well, it made me drunk. I think there must be nicotine in it, because it was aiming at substituting for cigarettes. Some of the old miners would chew 'baccy. You've seen Clint Eastwood, spitting in the spittoon? They'd spit out, deadly.'

Joe White worked at South Leicester, Ellistown, Bagworth and then at Asfordby. He said, 'One character I remember was Teddy Allen. I found out that Teddy was quite well known in the Leicester area for match fishing. Down the pit, Teddy used to sit there when we were taking breaks, and he was always shaving balsa wood, making floats.

'Another was Billy Black. I think for young miners like myself in South Leicester and Ellistown, we were working with a highly skilled workforce who were all characters in themselves. Some of the things Billy used to say. One example – I used to travel to work with him, because at that time I didn't have a car, and we was working a shift that was out of the scheme for the buses. So Billy would take me to work, and we'd pick a guy up, Albert Low, at the top of Whitwick. I remember one day, a lovely summer day, it was really hot in the car, I couldn't breathe. Albert had got the window down and he suddenly said to me, "There's something wrong with this car. The cool thing's not working right." And when I actually looked down he'd got the heater on full blast!'

Many of the miners were given nicknames, sometimes based on their surnames or even inherited from their father. Malcolm Tudor informed me, 'There were a lot of nicknames down the pit. There was a lad at Merry Lees and he was called Benny Webster. His name was really Brian, but his father was named Benny. There was another fellow at Merry Lees, I don't know what his name was, but we always called him Bronco. I think it was because he walked like a cowboy who'd just got off his horse.'

'Ozzy' Osborne (his real name is Cecil, which horrified his wife when she found out) tells me that he was the original Ozzy Osborne, long before the rock-and-roller. He also told me of another miner, Roger Wileman, who was known as Drop Dog,

Ellistown, 1957. (Michael Conibear)

87

Whitwick men: Frank Grady, Dennis Squires, Alan Pearce, Ralph Thorpe, Jim Bradford.

following an incident when he dropped his dominoes during a game.

Derrick Holmes recalled, 'There were a lot of nicknames. Mine was Decker, and my dad was called Jackety. There was Ken Richards – he was called Dutchy. He was a character, he was, at Desford.' Barrie Hall said, 'I had a nickname down the pit, I was called Brash. Well, my dad was called Brash, so I was Brash as well, obviously in them days. Everybody was called Brash or Tandy or Snarter and all these. It wasn't just family names. I don't know how they worked them out sometimes. There was even one lad down there and he were called Pancake for some reason.'

Not everyone appreciated their nickname, as Alan Ratcliffe told me. 'There was one pit bottom man, his nickname was Bacon. He didn't like it, didn't like being called Bacon. I know once a cage full of men were going up the pit and just as they got clear of the pit bottom, they thought that's it, and shouted, "Cheerio then, Bacon." And he stopped the cage and brought them back down again. He gave them some tongue pie.' Alan also told me about a specialist heading team that consisted of three men who sounded like they belonged on a farm: they were called Wasp, Fox and Cockerel. He added another nickname, a chap called Booty, so named because he sold clothes and could obtain anything his workmates wanted to buy.

Alan Pearson offered, 'My uncle that worked at Snibby, he were there 51 years, and they used to call him Tricket. In fact I had two uncles who both worked at the pit well over 50 years each, and they called the other one Chidder.'

Mick Richmond commented, 'When I was passed out as a machine driver at Bagworth, the first time I met Cliff Jeffrey – they call him Geek – he was a machine driver, a very, very good machine driver, and he always referred to people as old crust. "Ay up, old crust. I'm Geek. You're Richo, ain't you? What's your real name?" I said, "Mick Richmond." "Richo's better, isn't it," he says.'

Joe White recalled, 'There was quite a few nicknames. I remember two, son and dad, that worked at Bagworth. One was Arthur Moon and his son was Steve Moon. We used to called the dad Full-moon and the son Half-moon. There was one guy we use to call The Vest, he was a deputy. The reason we called him The Vest was because he was always on your back.'

Chapter 8

Tragedies and Some Near Misses

The possibility of injury, or even death, down the mine was an ever-present worry to the families of many Leicestershire miners. In some cases, this led to parents refusing to allow their sons to become miners when they left school. Alan Pearson told me about his grandfather, who was one of two men killed in a collapse at Whitwick pit in 1941. Because of his family's reluctance to see him go into the mine, Alan had to wait until he was a married man of 28, before he eventually went to work at Snibston Colliery.

A collier's death in December 1910 led to the award of the Edward Medal to one of his comrades. Charles Marshall and William Birch were working together in Coleorton mine, when a roof fall trapped the two of them. William Birch extricated himself but his workmate was trapped. As Birch tried to free him, there were three more falls that covered the two men. Throughout it all, William Birch disregarded his own safety and fought tirelessly to free the other man. Sadly, Charles Marshall died, but William Birch's courage was recognised with the award of the medal, known as the miners' George Cross. This was a rare event, and the only time it was awarded to a Leicestershire miner.

Derrick Holmes, who worked at Bagworth, Merry Lees and Desford collieries, told me of several serious accidents he'd seen while working down the pit, but he added, 'Thankfully, I was never there when anybody actually got killed. Although on one occasion, my mate, my pal, we'd gone out one night darting, and they come and fetched him out – we'd be about eighteen or nineteen at the time – and they said, "Your dad's got hurt at the pit." So I come out with him, and, as we walked down the corner, his uncle were coming towards us and he told him that his dad had actually been killed.'

Barrie Hall recalled another pit death. 'One day when we were training down Whitwick pit, we saw these lights coming towards us. There was four men carrying this, like, blanket. We said, "What's that?" and they really didn't want to tell us, but I reckon there was a body in it. It got a lot better though, through the Union.'

These were individual deaths, devastating enough for the families involved, but just over a hundred years ago the Leicestershire coalfield had its worst disaster. This was at Whitwick Colliery, when an underground fire broke out in the number 5 pit during the early hours of Tuesday, 19 April 1898. There were 42 men on the night shift, two of them working in the pit bottom near the shaft, plus the night shift deputy, Joseph Limb who was making his rounds. The remaining 39 men were working in stalls a mile and a half from the shaft bottom.

The fire had broken out in the roof timbers. Joseph Limb had made an inspection two hours before the shift started but noticed nothing amiss. Later, however, he first smelled and then saw smoke. He sent a boy, Albert Gee, to warn the men while he made further checks. On investigation he found that, at a spot some half a mile from where the men were working, the roof timbers were well alight and collapsing. After sending another man, Henry Springthorpe, to warn the trapped miners, the deputy made his way with some difficulty back through the smoke-filled return airway to the pit bottom. He sent one of the two men there, 77-year-old John Bird, a miner for 70 years, to the surface to raise the alarm. The second man, Charles Clamp decided to go further into the mine to warn the men and see if he could help to rescue them. This courageous act was to cost him his life. It was done completely on his own initiative, as the deputy thought he had gone to the surface with John Bird.

The alarm was raised at 4.30 am when John Bird ran to the nearby home of the colliery under manager, James Clamp. When the colliery whistle was sounded to raise the alarm, the wives and mothers of the trapped miners began to congregate at the pithead. Some were from Whitwick but others hurried in from neighbouring Coalville, Thringstone, Swannington and Griffydam. A number of clergymen, Nonconformist, Anglican and Catholic, also made their way to the waiting crowd to give what comfort and encouragement they could.

Hearses at the Whitwick pit disaster, 1898. (Michael Conibear)

Of the 40 trapped men, only five managed to fight their way out and survive. The other 35 perished. Efforts to extinguish the burning roof timbers failed, and the fire continued to spread. Over the next two days, brick firewalls were constructed to prevent air getting to the flames and the airflow was reversed, making it possible for a recovery party to re-enter the pit.

On Friday, 22 April, three days after the disaster, the first body was reached. It was that of the courageous Charles Clamp, the onsetter, who had gone from the relative safety of the pit bottom to try to rescue his fellow miners. Eight further bodies were found an hour later. They were in a group with their arms round one another. In the words of local poet Joseph Elson, they were 'Banded together when earning their bread, Banded together even when they were dead.' A new heading was made to reach the spot where the men had been working, and one body was recovered eight months later, in the December, but the bodies of a further thirteen men were not traced until January and February of the next year. Those of the remaining twelve, including that of John Tugby, a 16-year-old pony driver, were never found.

The ages of the dead miners ranged from thirteen to 63. In such a close-knit community, it is not surprising that many of the men were related. Samuel and William Stacey were cousins, while Henry and James Wyatt were brothers, as were James and William Davies. Although George Greasley was one of the survivors, he lost his father, William, and his elder brother Tom in the disaster. Joseph Tugby survived, but his younger brother did not. James Clamp, the under manager who had been roused from his bed at 4.30 am, was the father of the dead hero Charles Clamp. The tragedy at Whitwick Colliery left 28 widows and 94 fatherless children among the local mining community, and the loss is still remembered today, especially by the descendants of the men killed. Some of the widows, often with young families to support, were to remarry in later years but fourteen of them remained widows for the rest of their lives.

There were stories of lucky escapes. Tom Beniston had swapped from the night shift so that he could take his child to hospital the

Ellistown pit in 1961. (Michael Conibear)

next morning. Charles Pearce had also swapped shifts with a man that night, and survived. The man he swapped with died. Charles never got over the feeling of guilt that another man had died in his place. Sometimes the luck went the other way. John Davies died, having changed shifts with an old miner, telling him that he ought not to be working nights at his age. A number of men had gone off to Loughborough races, and thus were not at work when the disaster occurred. Ironically, the miners who belonged to the nonconformist churches would not go racing, and thus there was a large proportion of nonconformists among the men killed.

As well as this disaster, and other deaths down the pits, many miners had tales of near misses. Derrick Holmes told me, 'Until machine mining came, the conditions at Bagworth were pretty reasonable. But as soon as they started with machinery, they were exposing more ground and of course the conditions deteriorated. Plus the fact they were taking higher seams. The worst fall I've

Whitwick Colliery First Aid Team. (Whitwick Historical Group)

ever seen was at Bagworth. This particular morning, the overman told me, "I want you to go and make an advance road on Sixes Face. If you can just get a start, tonight we'll get a bit of dyno' and blow the way through. I want you to get ten to twenty rings in front, and you'll be paid heading price." So anyway, we went and by the Friday, we'd got about ten or fifteen rings in; they were about 4ft 6ins apart. We were in there, it'd be about 11 o'clock, near us snap time, and we'd cleaned up, and I turned round to Les and I said, "Crikey, look here." The thick iron bars were being squashed out of shape. I said, "Let's get out and get help."

'When we got to the end of the heading, everywhere shook. I'd been in the pit a long while but never felt anything like this. The props had tumbled everywhere and the roof had come in. The fall was higher than this house, you couldn't see the top. I looked down the face and I said, "There's somebody there down under the fall. He's a gonner, he can't possibly be alive." The next thing, this chap come crawling out. He must have crawled 20 yards or more. I couldn't believe he was all right. I'd never seen such a fall in my life. How he got through I don't know. The Lord must have pulled him through. He hadn't got a scratch.

'Then I had an experience when I was an overman at Desford. They were drawing off one of the faces, getting the chocks out, bringing them out on tubs. It was an atrocious road, low and 18 inches deep in sludge between the rails. You had to try to walk on the rails. We were bringing these chocks out, and the lads had got one jammed in a ring at the side. There was no room to get by. The deputy came along – a chap named Wilf Shepherd, who I'd been ratching with at Bagworth years ago – and he went up and got a bar to prise it out. Well, they got a bit of slack on it, and they let the chock come down and it pinned him.

'I was up another district, and of course, the tannoy went. "Decker," they said, "Wilf Shepherd's trapped. You better get there as soon as you can." So I went, but a shotman called Ron Ison had beaten me to it. What he did, he told the lads at the back to get a "sylvester" – a ratchet and chain device – to pull the tub back. And as soon as he got this moved back a bit, he jammed the bar in the floor, so he'd got it staked, and they pulled him out. I'm not kidding, his face was completely flat. We moved him down, out of the sludge, and then took him on a stretcher for 30 or 40 yards down to the main road. He was under the lights then, electric lights. Of course, we couldn't move him any more until the doctor come down to him. I think he administered morphine, something like that. They took him out of the pit and took him off in an ambulance.

'As soon as I got out, I had to go and report. I then had to go back down the pit, to take these chaps, the inspectors, and show them where it happened. Anyway, I come home, and I said to my wife, Blanche, "I don't know if he'll ever recover." But, do you know? He blooming well did. He was off work an awful long while, but he did go back down the pit. I really thought his face would be distorted but it wasn't. It just looked normal. That was the one of the most traumatic things I had to deal with.'

NCB ambulance at Lount colliery. (Chris Matchett)

Alan Ratcliffe, a mining surveyor, told me of one place where the roadway had been reduced in width because of stone piled up. 'Because they couldn't get the dirt and the stone out of the pit, they'd stacked it halfway in the side of the road. Stacked up neatly, but it did cut down the section of road. There was a tub track there, but the rest of the road was piled up with stone. I'd got my dial set up, and, as I was talking to a deputy, we heard this rumble back up the road, which was on the rise, sloping down our way. My first thoughts were that it was a pony running amok. But the deputy says, "Quick! Run!" Now when a deputy says run, you know it's something serious. It was a tub coming, a tub full of dirt.

'So I picked up my equipment – the legs with the dial on top – and ran in front of the deputy. It's not easy to run in the pit in between the sleepers. There were no manholes, nowhere to get in for shelter, you see. We'd just got to run it out. I wasn't getting on with this set of legs and the dial, so I chucked them both up on top of the waste stone. I mean, totally breaking the Coal Mines Act, that was. But as I chucked these things on the top there, they came off and they tripped up the deputy behind me. I heard such a clatter because hung on his belt he'd got his self-rescuer, the lamp battery, a detonator tin, the first aid tin and the safety lamp. There was such a noise, I thought, it's got him. Ernie Clitheroe, that was the bloke's name. Everything went quiet. I can't remember all the names he called me, when he did stand up.

'What had happened, this full tub of dirt was running amain – we used that expression, a "mainer", to mean it was running loose, running out of control. Another expression was the "doggy" or the "corporal". This was the roadman who'd be doing some roadwork, setting rails and sleepers or tidying the road up, and the name doggy came from the fact that they used road nails shaped roughly like a dog's head to fasten rails to the sleepers. On this occasion, the doggy had got a great big thick prop, and he chucked it across the road, behind that stack of dirt and the leg of the rings. And he stopped the tub, so it didn't catch up with us. But I was frightened. We both were.'

Alan also told me of another occasion when it was thought that men might have been lost down the pit. As a surveyor, Alan knew all the parts of the pit where many miners never went. 'In the coalfield, you know, you could get from Whitwick to Snibston, Snibston to South, South to Ellistown, Ellistown to Bagworth and Bagworth to Desford. All underground. Well, we went down one Whitsuntide, when the men were off for the holiday. We were doing a survey between Whitwick and Snibston, and it was a survey at the end of a face that had gone out about 700 yards, and then there was a knock-through into Snibston. You could get through into the Snibston road. We were helping to connect the two pits.

'We'd made arrangements when we went down Whitwick, telling the pit top man, "We shan't be coming out here today, we've made arrangements to come out at Snibston." But he never passed it on to anybody. And as it was a holiday, the deputies couldn't leave early because they couldn't account for the

Nailstone in the 1960s. (Michael Conibear)

surveyors. We hadn't come out. And one deputy had to traipse hundreds of yards up this district looking for us. He searched and we weren't there. Eventually he got to know we'd come out at Snibston, but he wasn't at all pleased. We came out about five o'clock at Snibby. A lovely red hot day, it was. We'd done the job while the pit was quiet, and we'd got a van waiting for us then, to take us back to the baths at Whitwick.

'I've seen that deputy many times later in Coalville precinct, and we stop and chat. I'd say, "Have you been up 100s District today, Jim?" "You bugger, you," he'll say.'

Chapter 9

On Strike

When the national miners' strikes of 1972 and 1974 took place, the miners of the Leicestershire coalfield supported them wholeheartedly.

Derrick Holmes' memories of the time recall how the Coal Board had made sure the miners would themselves run out of coal. 'In the 1972 strike, the Coal Board knew it was coming and they started delivering our concessionary coal later and later, so when the strike came, we had virtually none left. I used to go up the old sidings, where the rails had been taken up, but the sleepers were still there. I'd take a sleeper home and saw it up for fuel. I also used to scrat about for pieces of buried coal up there and bring that home. I found one piece as big as our sofa. My two lads brought their trolley up to move it, but when we put the piece of coal on, it broke the trolley. Still we managed to get it back home, and I smashed it up.'

Derek Howe was an active member of the National Union of Mineworkers (NUM), and he reminisces, 'In 1972, we hadn't got a clue how to strike. We'd never done it. The Transport and General Workers Union was organised to come along and teach us how to strike. Car workers knew how to strike, didn't they? So they came, and it was an early hours meeting, in the canteen. They told us a lot.

'We had to go up to Bakewell Street [the local NUM offices] and talk to the general secretary at the time, Flyer, they used to call him. He was a good chap, a Whitwick man. We went up there, and he told us how to sort ourselves out, put so-and-so in charge of a colliery, morning, noon and night. I got the job first, looking after the colliery. And the manager let us use the Union office, and that was very good of him. Someone – and I'm not going to say who – rang me and said, "Derek, someone's just broke through the fence to work, one of the foremen." So when the foreman went home, I met him and I said, "Now look, something could happen. If the lads get to know that you've been to work, they won't like it." So he never came again. We had a couple or three in, only surface foremen.

'That strike almost brought us to our knees, we weren't getting any money, but there was a deputy who did a morning milk round when he was on afternoon or night shift. And he used to leave us eggs and milk and lots of things for the blokes who were on strike. So we didn't do too bad in the '72 strike.

'In the '74 strike, I got the job of using my car. I used to take Albert Robinson, and we'd follow lorries to where they were going. Get the names of the lorries that were taking coal, against us. And we'd blackleg them when the strike was over. Well, one morning, we were somewhere going towards Birmingham, and these two lorries pulled in at this café, and so we pulled in too. We sat there talking, and this lorry driver came up and he said, "Are you two about ready then?" I looked at him, and he said, "Come on, I know you're following us." We followed them down the road, and they took us all the way to the Birmingham Gas Works in Saltley. We got on the phone – there weren't any mobiles then – to Frank Smith, and that day they organised a busload of pickets from our area to go there.

'I remember once, being up the Forest along with Albert. They told us up there, just by the Bull's Head, on the left-hand side in the hollow as you're going towards Copt Oak, there's a farmhouse. They said that the factories were storing coal there. So we went up, and me and Albert were looking over this wall, and all at once a gun came over. It were the farmer. Me and Albert never run so fast in our lives. We never did look to see if he was storing the coal.

'That strike wasn't as good as the first one, some of the lads got fed up and went in. The only thing that did upset us was there were a lot of chaps that left the colliery during the strike, and then, after the strike, they wanted to come back again. We didn't like that.'

By 1984, things were very different. Joe Gormley, a right-winger, had been replaced as NUM National President by the left-wing Arthur Scargill. When the Coal Board wanted to bring in a bonus scheme, the national NUM opposed it on the grounds of safety and also because it would mean that different areas would be receiving very different pay. Miners in difficult areas would be paid a lot less than miners where the coal was easier to get. There was a national ballot, and the majority voted against the bonus scheme. However, some areas went to court and got permission to ignore the ballot result. Leicestershire was one of the areas to adopt the bonus scheme. This may have been one reason why the NUM did not go for a national strike ballot when the Thatcher government began to close down mines in 1984.

The government had a number of reasons for wanting to close mines and, more importantly, inflict a defeat on the NUM. The earlier miners' strikes in the 1970s had brought down a Conservative government, and that was to be avenged. Also, a much reduced mining industry would be easier to privatise. Derrick Holmes told me, 'In 1984, Arthur Scargill was like a fish, because Margaret Thatcher hooked him. She built up stocks of coal at the power stations, then provoked him into calling the strike when she was ready. What he said was right, what he did was right, but he did it at the wrong time.'

The Leicestershire miners did not support the '84 strike. Some miners knew their pits were due for closure and wanted to earn as much as possible in the time that remained. One miner told me, 'There are the needy and there are the greedy, and I was one of the greedy.' Others, in spite of the bonus scheme ballot, said that they would not strike without a national ballot. Alan Pearson, a miner and a Labour councillor, told me, 'If we'd had a ballot in '84, believe me, whatever the result then I would have gone with it. But there was no way I would strike without a vote. I mean, in some respects what Scargill said has come true, nobody can deny that – but it was a principle. I honestly didn't think it was right. If he'd had the ballot, I would have supported the decision. I think once you lose them principles, you may as well chuck it up. I saw one of the Dirty Thirty standing on the picket line with a certain chappie, and I mentioned to him afterwards, I said, 'The nearest you've ever been to a bloody pit is when you've driven past one, so don't stand on a picket line shouting and bawling at me.' They'd got their principles, the same, though. They saw it from their perspective.

'It was difficult during the strike. At the end of the day I never felt right about it. I talked to my dad about it, because he could remember when his dad worked at the pit and they used to allow them to work two and a half or three days at a time, and then they couldn't sign on the dole for the other days of that week. That was before nationalisation. And of course they went coal picking when they'd finished their shift. My dad said, "Never let them get you like they got the 1926 miners." I always remembered that, so I was a bit torn.'

Not everyone was anti-pickets. Derrick Holmes explained, 'One day we had a lot of Welsh pickets come to Desford, where I was overman. They were very reasonable. I went out and explained that we had no young men and we weren't turning coal. We were just winding the pit down for its last few weeks. I said that the men would lose not only their pay but their pensions if they came out on strike.'

Joe White was another miner who felt torn between his conscience and the feeling down his pit at Ellistown. 'In the '84 strike, personally I totally believed in the fight to keep pits open. We were marginalized in Leicestershire really. Pits were on their last legs, but thankfully Leicestershire is an area where we've got other industries, got a choice of employment. I remember at the time actually feeling quite bitter and angry at the whole thing, bitter and angry at the government. Because if they closed

Bagworth, South Leicester, Nailstone, whatever pits was left in Leicestershire, we would find some employment somewhere, but then I was actually thinking of miners in the Durham, Yorkshire, South Wales coalfields, how difficult it was for them.

'We could all argue that we should have had a vote, it would have been different or it would not have been different. You could talk about that for ever more, but at the end of the day, that is history. If there had been a vote, I would have voted for the strike, without a doubt. I will say, in no uncertain terms, I do believe that every coalfield should have been out there the same as everyone else. I never crossed a picket line when Mick and the Dirty Thirty, or men from other areas, were picketing. But once again, we were marginalized in Ellistown. Bagworth was the mainstay of the striking miners, a whole group there came out. At Ellistown it was quite different. Maybe if I was at Bagworth at the time of the strike, I would have felt different and come out.'

Joe and Alan refer to the 'Dirty Thirty'. These were the 30 Leicestershire miners, out of a workforce of 2,500, who did support the 1984 strike. Originally called the Dirty Thirty as an insult, they decided to accept the title and use it with pride. Mick Richmond – Richo – was one of the leaders of the group, along with Malcolm Pinnegar, known as Benny. Richo told me how it began. 'When it was obvious the strike was going to come, I was up on Fours face. The Kent pickets had arrived. I came home and I spoke to this guy called Terry French. I said, "If I'm the only one here that's on strike, I will be."

'Well, the weekend came and went, and it was a crap weekend because I couldn't get it out my mind, you know, what was going off. I went back on the Monday, on day shift, and I said to the pickets, "I'll tell you what I'll do. It might sound stupid, but I'll tell as many as I can what your fight's about." I said, "Give me two weeks – that's a day shift and an afternoon shift – I'll get as many out as I can." And that was the agreement. But by the end of the second week, I was going mental, because it was all wrong, what we were doing. I was crossing a picket line, and I just couldn't do it. And that was that.

The banner of the 'Dirty Thirty'.

'And then I met all these other people, like Benny, Phil Smith, etc. It just snowballed. And that's when I heard this Radio Leicester interviewer, and she said to this guy, "Why have you called them the Dirty Thirty?" He said, "Well they're out of the coal pit, dirty, and there's only 30 of them, so the Dirty Thirty." And I thought, that's a good name. And that was it. In actual fact there were nearly 40 of us, so they could have called us the Naughty Forty.

'In the Dirty Thirty there was one from South Leicester pit, on his own, Johnnie Gamble. There were about 20 from Bagworth, one from South, then there was Billy Scott and Bobby Howard from Whitwick. Gordon Birkin, he was from Snibby. From Ellistown there was Martin Concannon and Keith Mellin. Bob MacSporran was from Power group, but he was out with us. Eventually we took some South Derbyshire miners in with us because there weren't many of them. So they came under our wing, but they had their own banner, their own badges, you know. There was 40 of us for a start, then a couple of them went back after nine months on strike.'

Although the Dirty Thirty received no support from the rest of the Leicestershire miners, other local people did help. Richo explained, 'Round here in Coalville, the railwaymen were heroes, they were. Absolute heroes. I remember little Ernie Hallam, bless him, he was a signalman. Brilliant guy. Out and out Socialist. And he would not signal coal. He was based at Bardon, on the signal box there. I used to go up and sit with him. I'd say, "When's the next train coming?" and he'd say, "Any minute. It won't get past." And it didn't. It would sit there until he finished his shift, and then one of his colleagues would come on, what he called scabs, and he'd signal it through. They stopped a lot of coal going through, the local railwaymen. I don't know how many millions of tons it was, but they were fantastic. Roy Butlin, Graham Cross, Dennis Wright, Phil Davies, Roy "Taggart" Clark, loads of them. Fantastic people.

'Amazing. We were just common miners, ordinary blokes and it had manifested into this. I could not believe it. We travelled about everywhere. We met these American people down in London, and eventually out of the blue they said, "We'd like you to come to Oberlin in Ohio." They paid for everything. Not everybody could go, and Benny didn't want to go. But about eight of us went across to New York – JFK Airport. I never thought I'd see that, never in my life. We stopped in Jersey City and some stopped in Brooklyn, and we met up every night. We were taken everywhere. One night we travelled overnight in a Greyhound bus – that was an experience – and we ended up in Oberlin College. I was roped in to being the English co-ordinator for our lot.

'I did a speech, giving greetings from the NUM, greetings from Leicestershire. I said, "I can't bring greetings from the scabs." And that brought the house down. It broke me up a bit, because I was so angry. They looked after us incredibly well there, they really did. This happened in 1985. The United Mineworkers of America, that's who did it. I was always amazed by these people where we stopped, and the places we went.'

The Dirty Thirty had a very difficult year. They were often out until late at night, giving talks, raising funds, getting into bed at 2 am. The next morning they were up early, organising a picket at the Leicestershire pits. Their working colleagues and some of their neighbours gave them a hard time. Malcolm received anonymous phone calls threatening harm to his family and even his life. One man – a working miner – broke into Malcolm's house one night and threatened to beat him up. Malcolm was not intimidated, however. He pushed him to the ground and sat on him until the police arrived to arrest the intruder.

Even in the worst of times, there was some fun. Richo recalled, 'Me and Benny, we always knew that our phones were tapped. You could hear it. I said to Benny, "There's only me and you going to know about this. I'm going to phone you later on, about seven o'clock, and I'm going to tell you that the North Derbyshire pickets are coming down and they're coming to Whitwick pit. But they're not. We'll just see what happens." He says, "I'm with you." And I phoned him and gave him that message, and then I went up to a phone box, near the Leicester hotel. I didn't make a phone call, I just pretended to. I saw one or two police vans, but when I walked down Whitwick Road, you've never seen anything like it. It was completely blocked with police cars and vans. I went back up to that phone box and I phoned Benny, and I says, "I wish you were here," because he was living in Stoney Stanton. I said, "It's worked. It's full of police." He said, "You're joking." I said, "I'm not." So that proved it conclusively.

'At the Jolly Colliers pub, there's a porchway and there used to be a swinging sign with a

Another jolly collier.
(Chris Matchett)

107

smiling miner on it. Peter Evans and his men had come up from South Wales, and they were getting picket money. Well, we didn't get picket money, so Peter Evans and Mike Thomas and one or two others said, "Are we going for a pint then?" and we said, "Well, we don't do that because we've got no money." They said, "Come on, we'll go for a pint." And Benny says, "Do you know where we should go? We should go to the Jolly Collier." We were in the pub and Mike said, "Did you notice that grinning bastard as we came in? The miner on that pub sign? I'll see you in a minute." And he took sheets of "Coal Not Dole" stickers, somehow got onto the porch, stuck a sticker on the miner's lamp, and others all round the collier's head. We didn't know what he'd been doing. But when we came out at the end of the night, he looked up at the sign and said, "See you then, bud." When we looked up there was this collier covered in "Coal Not Dole" stickers. Fantastic!

'Do you know it cost six million quid to police the coal strike? Six million. Absolutely unbelievable. A total waste of money. Some of those policemen made a fortune in overtime, just to police the 30 of us.'

The end of the strike came as a severe blow to the Dirty Thirty. In striking areas, the men marched back with pride behind their union banners, but in Leicestershire it was different. Richo described how bad he felt. 'Although in most areas, the NUM marched back to work with bands, and with their heads held high, we just turned up and walked in. I felt desolate, physically sick. Some people spoke to us, others didn't. I just eyeballed them. There was a hell of an atmosphere. The management wouldn't let those of us who'd been on strike work together. I was a machine driver, Benny was in the heads. The atmosphere definitely got worse as time went on.

'Then we got a letter, Phil Smith, Sammy Girvan and me, summoning us to one of the under managers. He said, "Er, well, it's not a punishment but we've got to be seen to be doing something. So we're going to send you on nights to clean out some drifts. The screens at Nailstone." So I looked at him and I thought he must be daft or something, this bloke. I said, "What have we got to do over there?" He said, "I want you to clean the drifts up."

'Now it was cold and wet, but we were put on regular nights, so we were getting 50% more wages than we were at Bagworth. We were

Desford colliery in the 1960s. (Michael Conibear)

stacks better off. And we weren't doing anything. We were just telling jokes most nights, and laughing us heads off. Wandering round Nailstone drift. And that was deemed as a punishment. We had three months there, before he finally dropped on that it wasn't a punishment at all, we were just making money for nothing. It was brilliant. We used to sit under the main drift belt that was whizzing over us heads, bringing the coal out from Bagworth. And we just sat, getting paid. All we had to do was clean the spillage up, off the belts. Three of us, all face trained. We loved it. And that was a punishment? Whatever.

'There was a lot of animosity from the other miners. But the one man from South who joined the Dirty Thirty, Johnny Gamble, he was really respected for what he did. There was a few men who were on our side: Terry Tracey, Cos Cooper. Cos has always said to me – he's a big friend of mine – he's always said, "You were right. I just wish I'd had that little bit more courage to have joined you." Some of my ex-workmates speak, some don't. Some even joined the UDM [Union of Democratic Mineworkers]. I speak to everyone. You forgive. But I will never, ever forget.'

Chapter 10

The Humour of the Pit

Miners developed their own sense of humour, a way of keeping their inherently dangerous working conditions in proportion. Policemen and paramedics are noted for their 'gallows humour', their dark witticisms and non-'pc' jokes that help them cope with traumatic situations that would have the rest of us in tears or on the edge of a breakdown. Colliers are the same.

The miners' sense of humour could make them see the funny possibilities of the objects around them. Derrick Holmes described how one miner at Bagworth pit, a chap named Tracker Siddons, raised a laugh as the miners on night shift were waiting in the pit bottom for the cage to take them to the surface. 'The men used to come into the pit bottom in the morning and queue up to come out the pit. Well, the first coal-loading machinery was going down the pit, and some of these pistons that were on the machinery, they'd got what looked like big straw hats on, for want of a better description. Padding that looked like a big straw hat. I can remember this particular morning, all of a sudden, it was all

Whitwick miners with the manrider. (Whitwick Historical Group)

Tracker had got on. He'd took this straw thing off the piston, and put it round his waist and he come down the pit bottom doing a Hawaiian dance. And that straw hat was all he'd got on.'

Taking the mickey out of the deputy was another source of fun. Derrick continued, 'And there was an old man on the road end switch, his name was David White, and what a character he was. He'd been a right old card in his day. Once when the deputy was phoning, David was at the road end. The belt had stopped, and the deputy was ringing up to see why. The phone was ringing and ringing and ringing. In the end, Davy picked it up and the deputy said, "Hello, hello, hello. Is that you, David?" David just says, "Hell-bloody-oh. It is David. What do you want?" The deputy said, "What's up with the belt?" David just said, "It's stopped," and put the phone down.'

It was not only the deputies who were the butt of the pranks. Alan Ratcliffe told me of an occasion when the joke was on a union official. 'Bagworth had had a new office block, including toilets. And, at one stage, the toilets had a blockage. They'd got the visiting union official from Ellistown, come up to Bagworth, and he was a very unpopular chap. He really thought a lot of himself. Nobody told him about the problem. And he was sat on the throne, not knowing there'd been any blockage. And Dyno-Rod had got to work, to force the blockage through in some way, with compressed air, I think. All of a sudden, there was an explosion underneath him on the throne, and it blew him right off the seat. All the stuff they were clearing up came out and went all over him. He was covered in confusion as well as in you-know-what. He was really in the mire. That story went round like wildfire. It spread through the area. "Did you hear about so-and-so?" And because he was so unpopular, it was a very enjoyable tale.'

Frank Gregory, who worked at Desford, told me, 'We had some laughs, of course. There were some of the most outrageous jokes under the sun, you couldn't even imagine what they were like, I mean, usually rude. Which I didn't really go for a lot, but you laughed with them. The blokes were normally working, they were too busy, but most of the jokes were at snap time and going up or down in the cage. When you were hanging onto the cage, hanging onto the bar, everybody were telling jokes. It was surprising how many, and it was really good fun.'

One of Derek Howe's jobs was to arrange for visitors to go down the pit. Where exactly he took them depended on who they were. On one occasion, there were visitors from Hobart House, the Coal Board headquarters in London. 'Once we'd had Hobart House down at Snibston, and Mr Ezra's secretary was with them, so we thought we would take them where there were no man-riders, so they had to walk, and where it was very sludgy. We took them right out and when we went to pick them up, the secretary says, "When I get back, I'll tell Mr Ezra what conditions you have at Snibston." It was good to take someone from down London and show 'em not the best – you shouldn't show 'em the best – but to show them exactly how it can sometimes be.

'I remember taking the Minister of Fuel down, a Conservative, I've forgotten his name but he was a director of Rowntrees chocolate. I told the lads, I said, "Now come on, you know what to do. You chockers, shift one or two pretty quick, and sharp, so that the back end comes down." That roof did fall with a clatter. It even frightened me. He couldn't get out fast enough. I shouldn't have done it really. He was all right, though. It was nice of him to come to Snibston, to see how the men worked. We took him to this face just to show him what it could be like. The highest part was a yard, not very high.'

Sometimes it was the visitors who terrified the miners. Derek recalled, 'We had this lady visitor, and she wanted to interview the miners when they came out the pit. Management had said okay, she could do that. But she went into the baths. And she says, "I'm not bothered about the men. I can interview them while they're showering." She might not have been bothered but the lads were. They said no way, so we stopped it. She only wanted to go in and talk to them while they were showering!'

Some lady visitors were more than welcome. Derek recalled another occasion when he took a party of ten American girls

Chocks supporting the roof at Whitwick Colliery. (Whitwick Historical Group)

down, girls aged about seventeen or eighteen. 'I took them onto the Yard Seam, because we'd got all young lads on there, all young chaps, eighteen to 25. We were going along the roadway, when they switched the speakers on. I'll be quite honest, the lads were all swearing – you never heard anything like it. I felt so embarrassed for the visitors. However, when we got on the coalface, it was all, "Hello, darling!" They'd only been out drinking with them the night before, most of the lads. I couldn't believe it. They all knew 'em.'

Derek also told me how one man used a practical joke to teach others a lesson. 'I remember one chap. The borer always came on at nine o'clock. When the chaps had cut some coal he could start boring holes. When he'd gone up the face a bit, these chaps used to drink his orangeade, you know. So he'd thought he'd steady 'em, so one day he filled it with urine. And the first one who drank it never told the others, because he was an old soldier who'd been in the jungle in Burma. Tougher than tough, you know. Never said anything, he just let the other two drink it. The borer had done it to stop 'em pinching his orangeade. All these things happened.'

This story of literally taking the proverbial was not just restricted to Snibby pit. Paul Liversuch worked at Donisthorpe, a colliery in Leicestershire, though always regarded as part of the South Derbyshire coalfield. Paul told me a story that illustrated how pit humour could be used to sort out a colleague who was not playing by the unwritten rules of comradeship. In the dusty atmosphere of the pit, a drink of water was essential, and each miner had his own water bottle. Of course a miner would share it if his mate had run out, but helping yourself to someone else's water was not done. In Paul's pit, however, someone was pinching water from other people's bottles. Although they knew who the thief was, they could not catch him at it. Their solution was – as at Snibby – a fitting one. An old water bottle was found and was ceremoniously filled with urine. It was then left with the other snap bags and water cans. The culprit found it and obviously tried it, because suddenly the stealing of water ceased!

Bagworth, 1974. (Michael Conibear)

Not all examples of miners' humour took place underground. Derrick Holmes told me of two occasions when he was away on a course. Derrick became an overman and it was decided that these officials should have training in speaking skills. He was on one such course where the men had to give a short talk to their colleagues. 'Roy Weston, he sat on this stage, and you'd got a red light and a green light, and as soon as the green light started you'd got to speak. And as soon as the green light started to flash, you'd got to wind down and the red light would come on to finish. You had to speak for about three minutes, and it seemed an awful long while. And to add to the tension for the speaker, you'd got Coal Board female staff there, all in the audience.

'Roy says, "Well, I'm going to talk about peace. You go to work, you're on the face, and all day it's just clang, clang, clang, bang, bang. You come out, you go home and your wife's going natter, natter, natter." He says, "So now I'm going to let you have some peace for about three minutes." He sat there silent with his arms folded. Everyone was laughing, but of course the convenor went mad with him afterwards. It didn't bother Roy though.'

On the other occasion Derrick was a long way from the Leicestershire coalfield, on a course at Newcastle on Tyne. 'And the first day we were there, we sat round. The course convenor gave us this résumé about his life, and then said, "Now I'll get to know who you are, starting with Derrick Holmes." Of course I stood up, and one man went, "Oh Danny-bloody-boy!" I looked round and as soon as he spoke, I knew him. When I went down the pit first, I was pals with a lad named Dick Miller, and we'd sat side by side in school. Until he spoke, I didn't know him from Adam, because he was white haired. I hadn't seen him since I was 14, and now I was 48. I said, "Dick Miller," and he said, "Ah." Of course the convenor wanted to know why he'd said that. When we were at school, we were singing Danny Boy. I was yawning, and the teacher said, "Stop. Derrick Holmes, sing that on your own." So that's why Dick called me Oh Danny-bloody-boy.'

Mick Richmond, 'Richo' of the Dirty Thirty, had a humorous tale to tell that came from even further away, when he went with other members of the Dirty Thirty to Oberlin College in the USA. It still causes him embarrassment when he remembers it.

'When we were in Oberlin, they asked us to run different workshops. They were asking about life down the pits and all that stuff. Towards the end, at one of the very last workshops, we were

South Leicester headstock in 1986. (Michael Conibear)

all on this stage and there were about 500 people out in front. We all spoke a bit, Phil Smith, Nigel Jeffrey, Bunny Warren. Then I had to do the last question and answer session, and this woman got up and she asked me, as I heard it, what I thought about the gays and the lesbians. I was going on and on, saying that I wasn't bothered whether they were gay or lesbian, they were all supporters and they were just fantastic people. I was going on about this, and at the end of it all, she came up to me and she said, "Actually, I asked you about the gains and lessons that had been learned from the strike." I said, "You're joking!" I died a death there on the spot. She'd said *gains and lessons* and I'd heard *gays and lesbians.*'

Chapter 11

The End of an Era

The winding wheel from Whitwick Colliery now stands as a monument by a stretch of road known as City of Three Waters. (They have unusual street names in the village of Whitwick, including another one called City of Dan.) A notice states that the winding wheel was erected on the initiative of Whitwick Historical Group 'to acknowledge the contribution made by the mining and quarrying industries to the local economy and to serve as a reminder of those killed during these operations'. On the old colliery site, what used to be Whitwick pit, there is now a supermarket with a number of other retail outfits.

Frank Gregory, who worked at Desford pit and still lives in Bagworth, recalled, 'When Bagworth Colliery shut in February 1991, the site looked a mess. The years passed, but they didn't touch anything at Bagworth. We felt really left out. It stood derelict until 1994, before we could get anybody to do anything for us. Anyway, when they did start doing things up, they did it in a big way. The Forestry Commission were involved. The spoil heap has been rounded and topsoil put on it. Then they started

The wheel from Whitwick pit, now a monument in the village.

planting 20,000 trees, and that's the pit bank, where the country park is. There's a wood called Tigers' Wood, after the Leicester Tigers' Regiment – they bought some ground and they've got a monument in the field there. The pit itself is now full of houses, because the coal board could make some money out of selling it for houses, see.

'Where Desford pit was, they made a really wonderful job of that. I can't believe it. When I go down there now, in my mind's eye I can see the old buildings that once stood there. But anyone seeing it for the first time wouldn't know there was ever anything there. Now, there's about four small lakes. One lake's on a higher level, and a stream runs in feeding it, so that it's nice and clear. Local people have made it into a fishing lake. There are swans – the old wildlife has come back, you can't believe how it's been. And that's a place where I spent 27 years working, and 27 years is nothing compared to some pitmen. The spoil heap at Desford, and that was a big one, it's nicely graded and there's walks over there, it's got signposts up and all sorts. They really made a good job of it.'

Miner's statue in Coalville.

The site at Snibston pit, in the centre of Coalville, was turned into a museum. This contains the old winding house and boiler house, shaft headgear, pithead baths and a signal box that used to stand by the level crossing in Coalville High Street. The tours round Snibston are all led by ex-miners. As well as mining history, the museum has a Fashion Gallery and 90 interactive exhibits. Outside there is a 100-acre country park and nature reserve. There is also a theatre within the grounds, which doubles as a cinema.

And what happened to the miners whose memories are recorded in this book? Some took early retirement, although their compensation was not the huge sum that people might suppose. Frank Gregory pointed out, 'The Coal Board were appalling. They didn't want to compensate you for anything, they'd give you the minimum. I was 52 when I finished and I got just £20,000.'

Malcolm Tudor left the mining industry in 1968 – 'I could see the industry was shrinking' – and moved to Warwickshire as a works electrical engineer at a cement works. As he put it: 'There was still dust around but it was a different colour – white instead of black.' After he retired, Malcolm and his wife Doreen moved back to north-west Leicestershire. Barrie Hall spent his compensation money on setting up his own driving school, and until he retired the Barrie Hall School of Motoring was a part of the local scene. Alan Pearson went to work for East Midlands Housing, where one of his first jobs was building houses on the site of the former Miners' Welfare building in Coalville.

Mick Richmond – 'Richo' – found it very difficult to find work locally because his fame as one of the leaders of the Dirty Thirty had preceded him. After the strike he went back to the pit until an ankle injury meant he had to finish. Then his problems started. 'I landed one job at the Tunnel, the Channel Tunnel. I passed the medical, I passed all the interviews, and they told me "With your coal cutting qualifications you'll be up the front in no time earning two grand a week." I said, "How much?" I thought my life's changing. But I had a letter come through one Saturday morning. I thought it might be a cheque for my expenses going down to Folkestone, but it was a note saying they couldn't offer me any work after all. And the top barrister Helena Kennedy – well one of her trainees, Isobella – she looked into it for me. She got back to me and said, "You're classed as a left-wing communist activist, and that's why you're not going to get a job."

'I kept trying for jobs and I heard about a firm up on Bardon

Industrial Estate. I filled the form in but I didn't hear anything about it for a long time. Then I asked one of the lads who worked there, I said, "Can you check if my application's in?" He says, "I'm going to do something, because I know I put it to the top last night." And when he went in the next morning, my form was at the bottom. Eventually he sussed out what was going on, and he told his manager. He says, "This application keeps getting moved, so your office isn't secure." Eventually I did get a job there. I had about ten years there, but I started losing time feeling ill, really ill. Eventually they found out that as well as having rheumatic fever as a kid, I'd had osteomyelitis, a bone marrow disease. And that can affect your nervous system. Apparently that had caused this chronic polyneuropathy, which I've now got. So – time is shortening.'

Richo's co-leader of the Dirty Thirty was Malcolm Pinnegar. After the Leicestershire pits closed down, Malcolm went to Coventry pit until that too closed. He had gained so much experience in fundraising during the strike that he got a job as a fundraiser at Leicester City Football Club. A keen Foxes supporter, Malcolm told me one thing that used to wind him up when Leicester City played in Sheffield. The Sheffield supporters would chant songs about the Leicester fans all being scabs – a reference to the 1984 strike, of course. For Malcolm, a man who'd risked everything by striking for a whole year, that was particularly galling.

Joe White had worked at Ellistown and Bagworth, then moved to the new Asfordby pit. When that proved a failure, he found a job in a very different milieu. 'When the pit finished, what do you do when you've had 21 years as a miner, and you're still only a young man of 38 or so? Anyway, I'd always had an inkling, it stems back from my trade union and socialist views, to work with teenagers. I saw a job advertised in the *Leicester Mercury*, I remember it very well. It said, "Are you empathetic, could you diffuse difficult situations?" And I thought, well yes, I could do all that. I've done that at the pit. Anyway, I put an application forward, never thought I got a chance, but I started in residential

The statue of a crouching miner at Bagworth.

services with Leicestershire County Council, in a children's home in Birstall. I still fervently believe that the closure of the Leicestershire coalfield was outrageous and diabolical, especially Asfordby. But I'm now working in something I really value and really enjoy.'

As well as the winding wheel in Whitwick, other local monuments to mining include a statue of a standing miner in Coalville town centre, and a figure of a crouching miner in the village of Bagworth. Frank Gregory, whose memories from Desford Colliery form part of this book, was asked to write a memorial. He told me, 'There's a statue in Bagworth to the miners, a bronze statue. It used to be outside the community centre until they found a permanent place on the roadside. We had collections for six years and it cost quite a lot of money – in the thousands. It had to be well anchored to the ground, else somebody would steal it. What pleased me, it made me a bit proud, a bit big-headed if you like, they asked me to write a commemoration to the miners to put on the plaque. I did it, and it's inscribed just as I told them, with my name at the bottom.'

Under the title 'Commemoration: Bagworth Mine 1825-1991: Desford Mine 1901-1984', Frank added the words:

'In grateful memory of all the miners who worked in the Bagworth Coalfield, some giving their lives. One of the oldest and longest running deep mines in the country, Bagworth Colliery was opened by Lord Maynard in 1825, and was the last to close in the Leicestershire Coalfield. The miners of Bagworth were the salt of the earth, courageous and very resourceful. They had a unique brand of humour and comradeship born from risking life and limb daily. In fact, they were guardians of each other's safety. During World War II they answered the nation's call for fuel with vigour, greatly helping the war effort. The gallant miners of Bagworth even entered the *Guinness Book of Records* with the following announcement on 5 April 1975: "The first coal mine in Britain to achieve more

Pithead at sunset. (Chris Matchett)

than 5 tonnes per manshift for a whole year." A truly remarkable breed; the miners of Bagworth can be remembered with pride.'

'A truly remarkable breed' – Frank's words are an apt ending to this look at the memories of the men who worked in the collieries of the Leicestershire coalfield.

Bibliography

Baker, Denis, *Coalville – the First 75 Years* (Leics Libraries & Information Services, 1983)

Baker, Denis, *Tragedy at Califat* (Swanington Heritage Trust, 2002)

Carswell, Jeanne, & Roberts, Tracey, *Getting the Coal* (Mantle Oral History, 1992)

Eagar, A. E., *Coleorton and the Beaumonts* (Edgar Bachus, 1949)

Gregory, Anthony, *Life on the Leicester Line* (Anthony Gregory, 2002)

Guide to the Coalfields (Colliery Guardian, 1957)

Hale, Leslie, Colledge, John, & Wileman, Michael, *Banded Together* (Whitwick Historical Group, 1997)

Hale, Leslie, & Wileman, Michael, *Banded Together, part 2* (Whitwick Historical Group, 2004)

Knight, J., *A Brief History of Whitwick Colliery* (Whitwick Historic Group)

Owen, Colin, *The Leicestershire and South Derbyshire Coalfield 1200–1900* (Moorland Publishing Co. Ltd., 1984)

Snibston Colliery 1832–1983 (NCB South Midlands Area)

Woodward, Maurice, *A Coalville Miner's Story* (Alan Sutton Publishing, 1993)

Index